The Militarization of the Western World

The Militarization of the Western World

EDITED BY

John R. Gillis

RUTGERS UNIVERSITY PRESS
New Brunswick and London

Library of Congress Cataloging-in-Publication Data

The Militarization of the Western world/edited
by John R. Gillis.
p. cm.
Includes index.
ISBN 0-8135-1449-5 (cloth) ISBN 0-8135-1450-9 (pbk.)
1. Militarism—History—19th century. 2. Militarism—History—20th century. 3. World politics—19th century. 4. World politics—20th century. I. Gillis, John R., 1939–
UA10.M5915 1989
355'.0213'09—dc19 88–36808
CIP

British Cataloging-in-Publication information available

Manufactured in the United States of America

Contents

Acknowledgments

I wish to thank all the contributors to this volume, as well as Terence Ranger and Andrew Pierre, who participated in the 1986 lecture series "The Militarization of the World, 1870–1986." That series was made possible by the generous support of the Humanities Grant Program of the New Jersey Department of Higher Education. Much is owed to Dr. Frederic Kreisler, the godfather of the program and a good friend. Among the many colleagues who were particularly helpful in organizing the series and who were especially encouraging concerning the publication were John Chambers, William O'Neill, Allen Howard, Lloyd Gardner, Victoria de Grazia, and Jim Reed. I also owe much to Marlie Wasserman for her support and guidance at each stage of this project, and to Cynthia Halpern for her expert editing.

Notes on the Contributors

GEOFFREY BEST has been Professor of History at Sussex University and the London School of Economics. He is author of *Humanity in Warfare* and *Nuremberg and After.*

GORDON A. CRAIG is Wallace Sterling Professor of the Humanities Emeritus at Stanford University. His many books include *Germany, 1866–1945* and *The Germans.*

CYNTHIA ENLOE is Professor of Government at Clark University. Her most recent book is *Does Khaki Become You? The Militarization of Women's Lives.*

MICHAEL GEYER is Professor of History at the University of Chicago. His books include *Deutsche Rüstungspolitik, 1860–1980.*

JOHN GILLIS is a Professor of History at Rutgers University. His most recent book is *For Better, For Worse: British Marriages, 1600 to the Present.*

PETER KARSTEN is Professor of History at the University of Pittsburgh. He has written many studies of the American military and is author of *Patriotic Heroes in England and America: Political Symbolism and Changing Values over Three Centuries.*

MICHAEL T. KLARE is the Five College Professor of Peace and World Security Studies, a program located at Hampshire College. Among his many studies is *American Arms Supermarket*, published in 1985.

PAUL KOISTINEN is Professor of History at California State University, Northridge. His *The Military-Industrial Complex: An Historical Perspective* appeared in 1980.

Abbreviations

ABM	Antiballistic Missile
AOA	Army Ordnance Association
ARVN	Army of the Republic of Vietnam
CIA	Central Intelligence Agency
CND	Campaign for Nuclear Disarmament
DDR	German Democratic Republic (East Germany)
EDC	European Defense Community
FRELIMO	Mozambique Liberation Front
GAR	Grand Army of the Republic
GFR	German Federal Republic (West Germany)
ICBM	Intercontinental Ballistic Missile
INF	Intermediate-range Nuclear Forces
LIC	Low-Intensity Conflict
MAD	Mutual Assured Destruction
MIRV	Multiple Independently Targetable Reentry Vehicles
NSC	National Security Council
OASW	Office of Assistant Secretary of War
SDI	Strategic Defense Initiative
SEALs	Sea, Air, and Land Capacity (U.S. naval special forces)
SIOP	Single Integrated Operational Plan
SLBM	Submarine-launched Ballistic Missile
SPD	Social Democratic Party
USIA	United States Information Agency
WIB	War Industries Board

The Militarization of the Western World

Introduction

JOHN GILLIS

IN THIS AGE of intense specialization, the broader themes of historical development tend either to be neglected or to be presented in the form of that uniquely modern genre, the textbook. This collection represents an exception. It originated in a three-part series of lectures organized at Rutgers University in the spring of 1986 devoted to the exploration of the history of militarization, focusing on the ways economic, cultural, and psychological preparations for war have shaped societies in every part of the globe. The nine invited speakers were all distinguished experts in their respective fields. They faced the formidable task of making intelligible and interesting to a general audience of students and teachers the nature of militarization in their particular area of expertise. Their efforts were so extraordinarily successful that many of those who attended the conference urged that the papers be made available in published form. We are fortunate that, with two exceptions, all have been included in the present volume.

While not all the contributors would make a sharp distinction between militarization and militarism, it is useful to try to explain why the former is the focus of this volume. Militarism is the older concept, usually defined as either the dominance of the military over civilian authority, or, more generally, as the prevalence of warlike values in a society.[1] Militarization, on the other hand, does not imply the formal dominance of the military or the triumph of a particular ideology. Instead, it is defined here as "the contradictory and tense social process in which civil society organizes itself for the production of violence."[2]

Like so many terms forged in the hot flame of public debate, militarism carries a meaning that is distinctly political and overtly ideological. As Michael Geyer observes, it is a term that is always applied to the "other." It is a way of displacing responsibility and blame, and thus must be used with great caution. Studies of militarism have been principally the domain of political historians. Defined in such a way as to direct attention to formal politics, the concept of militarism obscures deeper social processes and mystifies trends that transcend national political boundaries. One of the purposes of this volume is to break down the artificial separation between social and political history, placing the question of militarization more squarely at the center of general historical interest.

There has been a distinct tendency to identify militarism with particular political and social formations (the Prussian Junkers, for example), these becoming the standards by which other polities and social groups are judged to be militaristic or not. But judging one's own society against this "other," even if it is an ideal type, begs the question of whether militarism itself might mean different things in different societies, and, furthermore, whether its definitive characteristics might not change over time. The apparent clarity of a term such as "warlike" must also be called into question. Given the quite different requirements of major wars before and after 1914, it can no longer be self-evident what actually constitutes a warlike society. In short, treating militarism as an ideal type inhibits rather than enhances our ability to understand the past as well as the present.

Militarization as a term is less burdened with specific national and chronological associations. For one thing, it is a much newer term, having come into academic currency only in the 1970s and 1980s.[3] Militarization is less a thing than a process, one that does not depend on precise definitions of warlike values or even on distinctions like civilian/military, which developed in the classical era of constitution-making during the eighteenth and nineteenth centuries, but which have less and less saliency in the contemporary world. Furthermore, the distinction between the military class and nonmilitary class, once so clearly drawn between the aristocracy and bourgeoisie, no longer holds. Today, the line between the professional soldier and the civilian is blurred; and even the gender of the military, once so clearly marked, has undergone radical transformation.

It is true that the idea of militarization is a little blurry around the edges, but this is not necessarily a bad thing if it allows us to probe the actual processes of historical development at a depth and complexity not previously achieved. As a matter of fact, it is precisely because militarization is a historical process, an ever-changing set of relationships between military and society, that it cannot be pinned down for precise definition or projected onto the "other," as if it did not apply to us.

Unlike militarism, which has tended to be seen (particularly by Anglo-American historians) as something exceptional, archaic, even exogenous to modern society, militarization carries no such evolutionary presuppositions. It is not necessarily leading us in any predetermined direction. As Michael Mann has recently pointed out, there has always been a strong tendency, particularly on the part of Marxists and liberals, to see militarism as an atavism, as something that would disappear with the passage of time.[4] However, while the regimes and institutions that progressives identified with militarism have been vanquished, no one would seriously claim that today's western society is less organized for violence than it was a century ago.

The concept of militarization allows us to explore the relationship between organization for violence and modernity in a more sophisticated manner. Because militarization takes so many different forms over time and space, we can compare the militarization of one society with that of another without making judgments as to which is more militarized. The old concept of militarism served to shift the blame to others and divert attention from a society's own condition. Militarization, on the other hand, should force us to take a long hard look at ourselves.

The concept of militarization also compels us to confront history in its totality and to override the conventional distinctions between political, economic, cultural, and social history that currently dominate the field. Traditionally, political historians have concentrated on activities of decision-making elites operating within legally constituted polities. This approach served well as long as the historical players operated according to the classical definitions of the political and the military. But in the twentieth century the rules of the game began to change, and political and military historians have not always responded accordingly. They have been slow to come to grips with the "politics" of the civilian and military bureaucracies, of interest groups, and of supersecret agencies, all those extraparliamentary and extraconstitutional activities that today often have a greater determinative power than cabinet meetings and congressional votes. Apparently self-evident distinctions like military/civilian, war/peace, militaristic/pacifist no longer have the same meaning they once did. Even the notion of the military no longer holds; and because the organization of violence has undergone such rapid internationalization, it is no longer clear that the nation is the proper unit of analysis. To those studying the military in the Third World, it has been evident for some time that the inherited concepts and language of western political theory do not comprehend twentieth-century realities.[5] The work that has been done on militarization in nonwestern countries now provides new ways of looking at European and American history.

But political and military historians are not the only ones who need to readjust their approaches. One might have thought that economic historians would be particularly interested in the rise of the military-industrial complex, but they have not been. The new social historians have been equally negligent.[6] While there have been good studies made of social change in wartime, the processes of peacetime militarization have been largely ignored. While there is a considerable literature on men at war, the historical study of masculinity and its relationship to militarization is in its very beginning stages.[7] Greater progress has been made in women's history, but, there too, the focus has been more on wartimes than on the more continuous processes of militarization.[8] Cultural historians have been particularly negligent, although there is now a growing literature on what Robin Luckham calls "armament culture."[9]

Traditional disciplinary boundaries reflect and reproduce the language of politics and society inherited from the nineteenth and early twentieth centuries. Today, these are being challenged by the realities of a world in which the distinctions between government and industry, private and public, civilian and military are in flux. Class, gender, and age have also ceased to be the natural things we once thought them to be, and they are now seen to be historical constructions. The current situation points us in the direction of attempting to see the interconnections that make up the overall process of militarization in any given society. We are now able to see more clearly the roots of those relations between economic development and military preparedness that culminated in what is known today as the military-industrial complex. The necessity of probing the history of class and gender relations is also clear. Nor is it possible to understand militarization without knowing something about the changing nature of age and generational relations over time. Culture and the nature of cultural production and reproduction can no longer be ignored. And while it is a truism that militarization is a function of the nature of the state and of relations between states, international relations can no longer be conceived of merely in terms of diplomacy and military engagement, but must be seen as a much more subtle system of exchanges and interactions in which the militarization processes of sovereign states are being constantly influenced by one another, often without any control or even any consciousness being exercised over the process. We live in a world in which the economic, social, and political processes have collapsed into one another and where state boundaries have become quite permeable.[10] In such a situation, only a multidisciplinary and multinational approach to militarization will suffice to reveal all its levels and complexities.

A collection of this kind cannot pretend to provide a total history of militarization, but it does provide inspiration for and insight into how such a project might be developed. It is clear from Geoffrey Best's and Peter Karsten's descriptions of pre-1914 Europe and America that the processes of contemporary militarization, especially the foundations of what is now known as the military-industrial complex, were already in place by the First World War. It is only through hindsight that we can now detect the origins of those interdependencies between the economy and military preparation that have become so salient and obvious in recent decades. On the other hand, Karsten and Best also call attention to one of the interesting discontinuities of modern militarization, namely the striking difference between the blatant jingoism of the turn of the century, and the virtual cessation of the glorification of war that has occurred since 1945. Fascination with things military has not entirely disappeared in the western world, but it has become, as Michael Mann has so aptly described it, "spectator-sport militarism," aggressive in its rhetoric but passive in its behavior.[11] In the most highly industrialized countries, warmongering as such has gone out of fashion.

It is clear that a major change has taken place, but, as both Best and Karsten are at pains to show, the glorification of war before 1914 was not necessarily what it appears to be. Karsten demonstrates how the militaristic metaphor was as prevalent among antiwar activists as it was among the aggressive nationalists; and Best finds that, despite all the brave talk, the actual appetite for battle was weaker than one might think. Despite the massive efforts made to infuse the masses with a desire for battle, the reasons why men and women lent themselves to the war effort were actually quite diverse and essentially defensive. As recent literature on the soldiers of the Great War has indicated, men fought for a whole range of reasons—personal honor, peer group loyalty, regional identity—which had little to do with much-publicized goals of militarism or nationalism per se.[12]

This raises all kinds of interesting questions about the old generalizations about militarism; it also calls into question the national frame of reference in which most history is still being written. We need to remind ourselves that the nation is a relatively recent invention and that the "nationalization" of the masses had made very slow progress prior to 1914. The greater part of the European peasantry and working classes still regarded the nation as an im-position and a burden at that time.[13] It could be said, as Best argues, that the middle classes of western Europe and America had been "nationalized" by 1900, but we must not confuse our responses to the symbols of blood and glory with the actual meaning that people then attached to those symbols. If there is one lesson to be learned from these studies, it is that what is said and what is done are two quite different things. This applies as much to the present as it does to the past.

Militarization is not a singular, unified process, moving lockstep throughout society. On the contrary, the articles by Koistinen and Geyer suggest that it is uneven and sectional, different nation by nation. Paul Koistinen's exploration of interwar America shows how militarization took place quietly, and, even when exposed by the Nye Committee, continued unabated. American militarization took place without fanfare and with little of the militaristic rhetoric that was typical of interwar Europe. While it could be said that the old militarism glorified war but often failed to prepare for it, the emerging militarization process intensified the preparation while concealing its purposes and obscuring its consequences.

In his examination of European militarization, Michael Geyer exposes the myth of militarism, which invariably defines the militarist as the "other" class, gender, or nation. He shows that civil society was not an innocent passive victim of the seductive power of military elites, but an active participant in a massive social reconstruction that began with the First World War and is still evolving today. Geyer refuses to assign to war itself an autonomous power to shape politics and society. Instead, he shows us how it provided the occasion for a bourgeois male social order, already in the throes of an international crisis, to reconstruct itself on an entirely new basis.

The Great War had transformed the relations between civilians and the military, between government and industry, between state and society in all countries, but nowhere more than in Germany. While the Germans continued to be perceived as the most "militaristic" of all European societies, Geyer shows that militarism is too simple a notion to use to comprehend what actually happened.

In the end, it was not the military or military values that triumphed. On the contrary, the German military was overwhelmed and destroyed by forces and events over which it had no control. European militarization could not be contained. It fed on itself, ultimately destroying those national units that had given it birth, in order to create an entirely new international order dominated by the superpowers.

Looked at from a conventional perspective, the era of the Great Wars (sometimes referred to as the New Thirty Years War) seems to be the most militaristic in modern times. Bracketed as it is between the relatively peaceful nineteenth century, with its many "little wars" but avoidance of global conflict, and the age of nuclear deterrence, when the great powers have managed to avoid direct engagement, it is tempting to see this as an exceptional period. However, Geyer makes it clear that the societal transformation that began in the 1914–18 war continues into our own times. Even if nuclear weapons had not been invented, militarization as we know it would have occurred.

It is easy to ignore the changes that have occurred if we use militarism as our only measure. In America and Western Europe civilian control of the military remains the norm. As Gordon Craig's essay illustrates, the military is no longer a formal presence in German social life. Even its ceremonial occasions are met with violent opposition. No one dares glorify war; for all intents and purposes the old militarism is dead. Even after the Reagan military build-up, the American armed forces remain out of sight if not out of mind. It is only in Eastern Europe and particularly in the Soviet Union that military display and rhetoric are still in fashion, and there they have become highly ritualized, formalistic, and defensive in nature.[14]

But, if we look beyond the usual indicators of militarism, it is not change but continuity that reveals itself. Rhetoric may have been demilitarized, but the economy has not. To take just one measure of the increasing dedication of resources to military purposes, consider the fact that today military research and development absorbs an estimated 20 to 25 percent of all the world's manpower and material resources devoted to research and development. Around the globe, about five hundred thousand highly skilled scientists and engineers devote their talents to organizing for violence.[15] The boundaries between government and industry, already made paper thin by two world wars, are today even more porous. As Cynthia Enloe's contribution makes very clear, large numbers of women and men who would recoil at the thought of wearing a uniform or submitting to military discipline are now just as much a part of the military establishment as are the actual soldiers. Notions of masculinity and femininity have been radically reshaped, making the old John Wayne stereotypes wholly out of date.

And there is the fascinating question of the degree to which military metaphors and military ways of doing things have become so much a part of our lives that we do not even recognize their origins. The nurse's uniform, the marching band, and the Salvation Army were all inspired by things military.[16] We have found no substitute for the military metaphor in such diverse fields as medicine (fighting cancer), social policy (The War Against Poverty), and, of course, sports in general. Civil society often pursues its goals with a militancy that would be condemned among military personnel. Metaphoric militarization sustains that which is best as well as that which is worst in all of us.

According to formalistic notions of civilian/military relations, the United States and Western Europe seem to be far less militarized than the Soviet Union and many Third World regimes. But this does not take into account the degree to which military bureaucracies have grown in size and influence during the nuclear age. As Paul Koistinen points out, the power of military advisers was already evident in the Roosevelt era; and Michael Klare argues persuasively that war and threats of war are very much at the heart of American domestic politics. Military advisers (like Oliver North), paramilitary units, covert agencies (many of them associated with the military) operate in a shadowy realm, often without the oversight of constituted legislative or executive authorities. Arms sales, which have become an increasingly large sector of world trade, now dictate rather than follow diplomatic strategy.[17] Even where there is no overt politicization of the military, there is an evident militarization of politics.

If judged by the old criteria of militarism, we live in a very unmilitaristic time. But if we use the measure of militarization, a very different reality emerges. We have managed to compartmentalize our lives and our thinking in such a way that the truth rarely intrudes. This is perhaps inevitable in the nuclear age, when war is literally unthinkable. But, as Michael Klare shows so clearly, "little wars" have been a constant preoccupation of contemporary American politics. Americans and Europeans have had plenty of experience with war since 1945, and it is clear that they would prefer to have others fight their battles. If the rhetoric of the nineteenth century was a good deal more bloodthirsty than our own, at least it was more honest. We have lost the desire to glorify war, but we have also lost the ability to confront its deadly consequences. Ironically, this is partly the result of the unprecedented affluence we have achieved since 1950. Before that time, people had always had to make the choice between guns and butter. Now, at least in the highly industrialized world, it seems they can have both, thus avoiding hard decisions about resources.

However, even the mighty economy of the United States failed to deliver both guns and butter in the 1970s; and in the far less optimistic times of the 1980s, the contradictions between economic growth and military spending are once again apparent. But it is precisely because militarization now runs so deep and so silently that we have such trouble preceiving and articulating the exact nature of this dilemma. Too much of our national attention becomes transfixed on the few remaining remnants of the old militarism, as exemplified by Rambo, while the cadres of the new militarization, the men and women of Livermore Labs and SDI, remain invisible. Our consciousness of militarization lags behind our concern for militarism. If this book can overcome that discrepancy and encourage further investigation of the long-term processes that have brought us to where we are today, it will have served a most useful purpose.

PART ONE
Militarization before 1914

Chapter 1
The Militarization of European Society, 1870–1914

GEOFFREY BEST

THE YEAR 1914 conjures up a number of images, familiar, effective, and not without some truth in them: Europe the armed camp becomes Europe at war; the end of an old order, a good old order; a long summer giving way to a winter that has lasted ever since.[1] The war began with extraordinary exaltation and rejoicing. The nations that plunged into war gave an appearance of doing so gladly and enthusiastically. The war that began was soon recognized, and has ever since been described, as a war by timetable. And within three or four weeks, the already large numbers of men who were under arms at the beginning of August had gone up to as many as seventeen million. Within the little over four years the war lasted, the number of men mobilized reached over sixty million, and, of that number, France, which contributed the highest proportion of its population to the war of any of the belligerents, had lost almost one-and-a-third million dead and had inherited over two million seriously wounded—the tragic *mutilés de guerre* who were to haunt the streets and parks of interwar Europe.[2]

The images and statistics would seem to suggest that a militarization of European society had happened, if not already by August 1914, then as soon as the hostilities had begun. There is some truth in this, but just how much is open to question. We can best determine the degree not by holding Europe up to some abstract model of militarism, but by cutting slices into European society before 1914 to examine four aspects of the process of militarization, to see how civil society was organizing itself for the production of violence. We will look first at industrialization, the mobilization of resources, material and human, for potential use in warfare. We will also examine class cultures to see how the rise in the standing and power of the middle class (bourgeoisification) affected militarization. In addition, it will be useful to examine what effect the emergence of the modern women's movement had on civil society's organization for violence. Finally, there is the question of the culture and mentality of war as that was developing nationally and internationally before 1914. As we shall see, militarization proceeded in a most uneven manner, differing from one nation to another, and it was productive of all kinds of contradictions and surprises within each society.

Without industrialization modern militarization is inconceivable. Yet, while it produced the sinews of war, industrialization also posed serious problems for the reproduction of modern soldiers. The same industrial cities that manufactured such mighty weaponry also produced political and social ideas that were uncongenial, to say the least, to military planners. The modern millions could not be dispensed with. From the point of view of the elite planners of organized national violence, the industrial working class was something of a mixed blessing: it was economically advantageous, but politically perilous; not at all what the old military aristocracy or new military professionals were used to dealing with or wanted to deal with. But they had to be dealt with because this was a Europe of competitive armaments, compulsory mobilizations, a state system that was clearly bent toward war and bound to live in constant preparedness for it and for the increasingly obsessive numbers game of comparative demography.

In Germany the problem was at its most acute because industrialization there was so intense, and Germany was of course the country in which the military aristocratic caste was most strongly entrenched into the constitution and the social order. How did the military managers of the Second Reich handle their German version of the problem common to all these countries—the problem of bringing into the national armed forces and into conformity with the national readiness to fight the growing industrial, urban-based working class whose political and cultural preferences seemed likely to go against those of the ruling classes?[3]

One essential aspect of their handling of this problem was the use they made of conscription. It became taken with a new seriousness. The 1870s were a pivotal time in the history of military service in Europe. The obligation to universal military service existed in every country, but more in theory than in practice before the seventies except in the German states. After the Franco-Prussian war, the whole continent moved in the German direction. In one country after another, legislation in the 1870s purported to make universal adult male military service something of a reality. But how much of a reality was it really? The public impression that states sought to give seems to have been a more formidable one than the actual realities located in the camps and the barracks. In the French army, for example, between 1905 and 1913 we discover that about half of the soldiers doing their second year of conscription were doing it in noncombatant jobs. In Russia, Norman Stone tells us, more than two-thirds of the possible conscripts were being exempted from the call-up. The tsarist state, it seems, simply could not afford to feed them all. As for Finland, that peculiar western province within the Russian empire, it seems that no more than four lads were conscripted in 1902. Even in well-organized and relatively rich Germany, about half of the possible conscripts were actually missing from the ranks. The cost of universal military service was, in fact, preventing anything like the full call-up that the legislation and national claims suggested. And that fact was only as well concealed as it was because states did not wish such disagreeable facts to be publicized, any more than patriots wished to talk about them.[4]

Conscription was also an opportunity for indoctrination. The conscripted millions were not called into the army simply to be trained for military service. They could also undergo training for political and social service. The army was, sensibly enough, perceived as a school for patriotism. This use of it was intensified in the more industrial countries in order to counteract what were supposed to be the bad influences of industrial urban civilization and socialism. Especially in the German army this was felt to be a problem, because in Germany the national elites felt more anxiety than anywhere else in Europe about the cohesion of their multipartite empire, containing as it did so many minorities and special interest groups. Germany's intense and successful industrialization from the 1860s onward brought with it the biggest socialist party and the biggest trade union movement on the continent. This industrial working class was chief among those *Reichsfeinde* identified by the ruling elites as sectors of society inimical to the great national purposes of their German Empire. And it was therefore judged peculiarly necessary to use the period of military service in order to counteract whatever antinational and unpatriotic inclinations working people might bring with them into the army and to reconcile them so far as possible with the established political and social order. German officers and to some extent noncommissioned officers spent a lot of time indoctrinating and fathering their charges; with what results, only the outbreak of war would show.

In France the problem was perceived in rather different terms. There, the question from the later seventies on was how to turn the adult manhood of France into a body of good republicans. And whereas in Germany, the army was seen directly as a school for imperial patriotism, in France it was of course seen more as a school of citizenship and of republicanism. This intensified at the turn of the twentieth century under the premiership of Waldeck-Rousseau and his war minister, General André. Douglas Porch tells us that the results of the intense indoctrination and education given to the troops were actually rather unhelpful to the military efficiency of the army. There wasn't time enough for all the military training that ideally was needed alongside the civic education that the soldiers were receiving, quite apart from such limitations on military training as were imposed by the shortage of cash.

How well, we might wonder, did this indoctrination work? The prime proof of the pudding, of course, was going to be in the eating, and that could only be among the experiences of the First World War. But there is another significant little bit of evidence about the success of conscription insofar as it was meant to be a popular and acceptable experience. It comes from its evident unpopularity, in at least some social sectors, in at least three countries, Austria-Hungary, Russia, and Italy. It is very clear that in those countries the prospects of conscription and the distasteful realities of military service were among the principal causes of emigration among the younger male population. To that fact, Alan Sked, of the London School of Economics, adds the results of inquiries into the relative rates of suicide in the armies of Europe. The suicide rate in the Austro-Hungarian army was a lot higher than in any of the others—a fact that Sked persuasively attributes to the extremely unpleasant experience that conscription proved to be to such of the young males of the Hapsburg Empire as couldn't avoid it.[5]

This brings us to the question of what we have learned to call the "military-industrial complex." Its beginnings lie in the years before 1914. It is not a question of the quantity of production of war materials. Quantity was there for sure before 1914. The armaments industries of Europe became huge in these years. One indication of that, for example, appears in Arthur Marder's reckoning that about one-sixth of the British workforce was doing something for the service of the Royal Navy. But military-industrial complex means something much more than the sheer quantity in production. It means something structural and in part political. It means a symbiosis and a mutual dependence of the military side of the state with its industrial structure and its political system, the political system that contains and secures the others. Two very good writers about the military organization of Europe in these years, Pearton and McNeill, both point to the United Kingdom in the 1880s as the country in that decade in which something recognizably like this essential military-industrial complex came to birth.[6]

The advance and intrusion of the middle classes into areas and layers of society hitherto dominated by the aristocracy—the process described here as bourgeoisification—had contradictory consequences for prewar militarization. In the years after 1870 there had been a considerable penetration of the European officer corps by members of the bourgeoisie. There had to be some, if only because—to go back to a point of which one can hardly make too much—Europe's armies had to be far, far larger than the limited number of hereditary aristocrats could possibly manage on their own. Bourgeoisification in varying forms was therefore to be expected. Middle- and even lower-class characters had penetrated the officer corps and those corps became more diluted than they had been earlier, except where they were already well-diluted as, for example, in Spain and Belgium and Italy. We should also remember that the navies of Europe were always more bourgeois in their social composition than the armies, which means that where navies were or became huge, as did the British and German ones, these were large and important outlets for bourgeois militarism and patriotism. But the case of the armies is the more important because they constituted so much the largest part of the armed forces. Karl Demeter has documented for Germany the increasing proportion of middle-class men in the officer corps. Less well-known are the developments in Austria-Hungary and Russia. It may come as a surprise to find that in these countries where the domination of the aristocracy seemed so complete the middle classes had made great inroads. In Austria-Hungary, it seems that by 1913 as many as 88 percent of the general staff was of bourgeois origin. But 60 percent and more of the generals of the army could be thus classified and even over 50 percent of the cavalry officers. (More remarkable still, perhaps, nearly 10 percent of the officer corps was Jewish; a proportion higher than that of Jews in the population as a whole.) In Russia, the army had long been a ladder of social mobility. By 1914 nearly one-half of the officers of the rank of colonel and below are described as either lower-middle-class or actually peasant in origin.[7]

But what weight should be attached to these figures as part of the history of militarization? Not, perhaps, as much as one might expect. One would expect this bourgeoisification to have had a big result if one accepted the conventional aristocratic prejudice that the bourgeoisie was a class by inclination and upbringing less fitted for war than an aristocracy that was supposed to have notions of honor, service, and sacrifice bred into its bones. What the aristocracies held against the bourgeoisie when they contemplated their entry into the ranks of the officer corps was their supposed preference for material and money values and their ignorance of the imperatives of the chivalric code of honor and of service to monarchs. It doesn't seem as if those aristocratic prejudices were justified. Bourgeois officers proved just as well able to give their lives for their country as aristocratic officers ever had been, and they drew their codes of honor, service, and sacrifice from other sources; in the English case, for instance, the so-called "public schools," or in the French case, the republican code of virtue already referred to. The undoubted bourgeoisification of the officer corps cannot be assumed to have made to the morale of Europe's armies the differences its aristocratic critics expected and alleged. The new officers brought their own, admittedly different, moral imperatives with them, or, as was the case in the German army, they gladly and enthusiastically adopted the aristocratic code they found already entrenched.

Organizing civil society for the production of violence necessarily involved women as well as men. This was proceeding in a variety of ways, but not necessarily evenly or unproblematically. It is easy enough to see how women were mobilized for war once hostilities had begun, but it is less easy to establish how this was happening in the decades before 1914. Here we can use the insights provided by recent historiography on women, which has shown the many subtle and indirect ways that gender fits into the processes of militarization. Cynthia Enloe has shown how patriarchical ideologies use even the most pacific notions of womankind to serve the purposes of militarism. The photographs of grim-faced, machine-gun-toting young women associated with various current Third World national liberation movements have made us aware of the ways in which images of women have long been used militarily, whether or not women do the actual fighting. On the other hand, we must also acknowledge the role played by women and the images of womanhood in peace movements.[8]

All that has to be taken into account in the investigation of women and militarization in our own times. But what shall we find if we go back into that almost unimaginably different-minded period before 1914? The indoctrination of females to be mothers of warriors goes back a very long way in European history. The culture was rich in its Greek and early Roman origins in examples of heroic, self-sacrificing and tough-minded mothers who say to their sons, "Return with your shield, or on it," meaning either alive and victorious or dead with honor and respect. *"Dulce et decorum est pro patria mori,"* the old republican Roman tag, was as predictably put on the lips of the mother as of the patriarch. In our own century, we have been made familiar with imperialist phrases like "Mothers of the Race," with honorable and tragic patriotic ones like France's *"merès de familles nombreuses,"* more recently still with the "hero mother of the Soviet Union," and the good German woman who provided "a baby a year for the *Führer*"—such images remind us of the importance of the mother in the European warrior-state.

How much of this was already present in Europe before 1914 is suggested by Anna Davin's study of the connections in Britain between the imperial impulse and the exaltation of motherhood. The mobilization of women as mothers took place in the context of the great debate about British "national efficiency," triggered by the discovery of the unhealthiness of the recruits for the Boer War. We cannot be sure just how women responded to such propaganda before or during the war, but it is worth noting instances of German as well as British women expressing what can only be described as Spartan-Roman sentiments. They wrote to newspapers about how glad they were to give their sons to the country; and many were active later in shaming boyfriends, husbands, and sons into volunteering for wartime service.[9]

Now for culture and mentality. This is very important, for the set of the mind toward war and violence, after all, is a determinant of behavior. I find little disagreement among historians that however ready the peoples of Europe already were for war by 1870—and clearly nationalism and patriotism were common values more widely shared than their "internationalist" opposites—that readiness intensified within the next few decades. Any survey of the culture and mentality of Europe in that epoch will discover a number of novel developments that were taken at the time to mean a heightened readiness for war and have been accepted as such by historians ever since. Of course we are left with huge problems of assessing the weight of "public opinion" and understanding how it affected public policy, a notoriously difficult subject.

Yet, it would seem that there was a growing military enthusiasm and commitment just before 1914. With regard to ideas and assumptions popularly held about states and their international relations, these years saw a fortification of the kind of ordinary defensive patriotism and nationalism inherited from the past. But we can also see the development of a sort of supernationalism or super-heated nationalism, the kind of thing typified for Germans by Treitschke and Frenchmen by Barrès, to name only two of a great host of popular patriotic and supernationalist writers. In Britain, it wasn't so much nationalism as imperialism that dominated the public mind. And to those nationalisms and imperialisms there were added in the same years new ingredients of social Darwinism and racism. All of this contributed to a climax of what Michael Howard and others of us call "bellicism," meaning, beyond militarism, a frank and even glad acceptance of war as a supreme experience of life for men who could undertake it, along with of course an absolutely unquestioning acceptance of war as an instrument of international politics. There was, for example, the reply Field Marshal von Moltke gave to the international lawyer Bluntschli when Bluntschli sent to Moltke in 1880 a gift copy of his new international law textbook, with its chapters on limitation and restraint in war, its tendency to deprecate unnecessary wars, and its clear preference for the peaceful resolution of international disputes. Moltke politely acknowledged the receipt of this book and said that he thought that not only was the hope of perpetual peace unrealistic, but that he feared that it was also distasteful and deplorable, because war was a necessary and healthy aspect of national experience. A world from which wars had been abolished was a world in which Moltke, for one, thought that humanity would be the poorer; nations that never had to fight would "lose their virility."

An interesting aspect of the cult of war in these years just before 1914 was the great enthusiasm that the militarists and supernationalists of Europe showed for the Japanese, just discovered in their successful war against Russia to be a nation of extreme warrior virtue and military efficiency. The Japanese readiness for national service and sacrifice was joyfully held out to the young of Europe as an example of pristine virtue from which the perhaps effete old world could learn. Finally, there is the 1910–11 edition of the *Encyclopedia Britannica,* which humanists and arts scholars have always been taught to regard as the best edition of that encyclopedia ever undertaken. So it may be, but cool examination of it shows that war and armies were given a quite extraordinary amount of attention and space![10]

To turn to literature—in these years we find not only the so-called "literature of action" based largely on the experiences of Europeans overseas, above all in Europe's empires, but also a new genre of literature, the "literature of the next war" so well described in I. F. Clarke's book, *Voices Prophesying War.* In more popular culture we find in these same years the new cheap press often marked by very strong jingoism, xenophobia, and trivialization. These characteristics are excellently taken off by Max Beerbohm in his cartoon showing a horrible little cockney journalist talking to Bellona, the goddess of war, a haggard harridan despondently standing there with her torch upside down. Says the cockney journalist to her; "Aw, come on Bellona, light up your torch. Britain and Germany has *got* to fight it out."[11]

The history of education shows the youth of Europe in these years being encouraged to join rather militarized or military-looking youth movements; the Boy Scouts, of course, being the prime example of these. And to some extent even girls and young women were being recruited into services and movements with some auxiliary military purpose, in connection particularly, as some might expect, with nursing. At the same time we find that history syllabuses in schools laid extraordinary emphasis on the place of wars and battles in history. Most popular history indeed was nothing else.[12]

In politics, apart from all the belligerent effects and influences we have already encountered, we find two special ones that point to the militarization of mentality and the increasing readiness for war. One of those features is the development, in every country where there was anything like a representative system, of nationalist and militarist pressure groups that worked more often outside the parliamentary system than inside it, with strongly national and patriotic or imperialist purposes. In Britain, there were bodies like the National Service League; in Germany, there were organizations like the Navy League, the *Flottenverein*. Another, related feature that marks these years and gives it a new character is the phenomenon of recurrent war scares from the later 1870s onward, war scares promoted by journalistic alarmism and fed by the xenophobia and the national competitiveness that was so marked a characteristic of public speech on international issues.[13]

But while all this would seem to have made the rush to war in 1914 predictable, there was also a different tendency at work on European culture that can be fairly be depicted as a "peace movement." It is as if the "war movement," as one is tempted to call it, had the "peace movement" in counterpoint. Although they would seem on first impression to be diametrically opposed, these movements actually sustained each another in complicated ways. There were degrees of attachment to both the war and peace movements. Just as there were pure bellicists heading up the war movement, so there were pure pacifists to be found in the peace movement. But in between those two extremes there was a wide variety of tendencies and a good deal of intermixture as well, as if the same mind and conscience could simultaneously or alternately contain both inclinations—the one toward violence and the other toward peacefulness. Why should we be surprised at such inconsistency in our ancestors of that epoch when we accept very much the same inconsistencies in ourselves and in our own times? The peace movement, like the war movement, was mixed; and two of its achievements in particular are worth a place in this sketch.

These ambiguities were especially evident in The Hague Peace and Disarmament Conferences of 1899 and 1907. Like all subsequent disarmament conferences, they were just as much armament conferences, and their most concrete achievement in international law and organization was the construction of a code of conduct of war rather than a set of institutions for the maintenance of peace (although something was done along that line too). But the main importance of these conferences is found in the fact that they happened at all. That they happened is largely because the chiefs of the states of Europe believed that the peace movement, with its many organizations and publications, was powerful and influential enough to deserve or to demand some recognition by the managers of state power that their conventional commitment to war preparations had now to be sweetened by some concessions and efforts toward the peaceful resolution of conflict. The two conferences at The Hague were politic concessions to a movement of opinion that was felt to be influential and sensible enough in many ways to demand it.[14]

Secondly, it has been persuasively put by the German historian Dülffer that, among the elites managing the Great Powers themselves, there was by 1899, and much more by 1907, a growing concern, not only about the expense of the war system to which they were all dedicated, but a nagging apprehension that it was going to issue in disaster of some sort; as if all but the most bellicist of the managers of power, whatever the firmness and truculence of their public language, experienced some growing disquiet in at least part of their minds about the war-bent course on which they were all set. That side of the peace movement, then, which issued in the great Hague Conferences, may be classified as its élite and establishment side. Another side, socially and politically very different, is to be found among the working-class organizations of Europe loosely linked in the Second International. This movement too pointed both ways. Just as the Hague Peace Conferences actually set up rules for the better conduct of war, so the Second International's rhetoric and reputation for seeking to prevent war through international working-class solidarity actually belied persistent differences between the several national parts of the movement and a good deal of primordial nationalism among its leaders. Notoriously the Second International's threat or promise to make war impossible by calling out the working classes against involvement in a capitalist or imperialist war with each other didn't work when it came to the crunch in the summer of 1914. This comes as no surprise to those of us who always thought that the emotions of patriotism and nationalism were stronger in the working class than the class loyalty claimed for it by Marxist dogmatists and plebeian rhetoricians. Now the historian Howorth has presented a mass of well-ordered evidence to show how both the French and the German labor leaders in the Second International were actually much more nationalistic in the parts of their consciousness and emotions where it mattered than they were internationalists in their public language and statements. There is no need to think that either posture was insincere. We need only find a way of explaining how internationalist aspirations and nationalist attachments could coexist in the same person and the same mind. It can be described as Howorth describes it as a "plurality of consciousness" and perhaps that explains much more than just the Second International.[15]

The First World War can be regarded as a test of the militarization that had taken place by 1914. At first glance the enthusiasm and acceptance with which European societies went to war would seem to be proof that militarization had proceeded in a straightforward, highly successful manner. First of all, consider that the seeming rush to war, the glad enthusiasm of those weeks of the later summer of 1914, was remembered by so many of the participants who survived them as the most exalting and exhilarating and encouraging days of their lives. It has to be said that governments were on the whole surprised by the extent of the enthusiasm for war. The French and German governments in particular had been quite apprehensive about the reactions of significant minorities of their populations. It was extremely important for the German government to present the war to its people as a war thoroughly justified, as a war purely and simply of national defense. Only if they could do that, the German government correctly believed, was there a reasonable chance of securing the patriotic loyalty of the greater part of their working classes. They succeeded in doing so. The French government likewise was relieved and surprised at the way in which its presentation of the war was accepted by the French people. The number of absentees from the call-up, which they had expected to be fairly large, turned out in the end to be fairly small.

However, this moment of enthusiasm was not necessarily a product of deeply held bellicist views. It could also be interpreted as a momentary aberration and not at all indicative of the more complicated attitudes toward war held by the vast majority of the population. For Britain, Keith Robbins has given us the following assessment of patriotic enthusiasm in the early days of August 1914:

> The crowds outside Buckingham Palace did chant "We want war," on August Bank Holiday. This apparently "natural" and spontaneous solidarity was not the result of any objective unsettling of social order and class distinction, but was more analogous to a carnival or fête involving a temporary suspension of social behavior and an indulgence in unproductive expenditure. The world was in a gorgeous state of flux.[16]

By that good historian's judgment, at any rate, the glad rush to war had origins and explanations much more casual and superficial than those of profound warlike enthusiasm and patriotic self-immolation that has too readily been supposed.

The French historian Jean-Jacques Becker, in his two fine books about how the French went into war in the first place and about French opinion and morale through successive phases of the war, has concluded that the bulk of the French people were not moved by the sort of ardent nationalism that Barrès and the hypernationalists had been peddling before the war broke out; but they were moved rather by good old, traditional, uncomplicated patriotism: that was the prime emotion that took them into the war and kept them going through its many awful months. Peter Simpkins of London's Imperial War Museum likewise has concluded from his intensive study of Kitchener's army that the ideas that principally moved the soldiers were, first, a genuine kind of moral feeling, the conviction that Belgium was a good cause for which to go to war, and second, a patriotic pride in empire; among those willingly belligerent hundreds of thousands he finds no particular aggressiveness toward Germany and certainly little relish of war for its own sake. What he does find is a profound British style of patriotism, transmuted, not surprisingly, into an imperialist mode. Of Russia, no need to say more here than that there was a good deal of working-class hostility to the first call-up—a number of nasty incidents at enlistment stations and so on, and the deaths or serious injuries of more than one hundred of the officials administering the call-up. But thereafter, although the desire to stop fighting and to get out of a wretched and apparently unsuccessful and endless war was very strong, primitive and basically defensive patriotism remained a strong emotion with the Russian soldiery, who, while no one could by 1917 mistake their desire to be out of the war, did not want to do it in an unpatriotic way.[17]

The question of persistence in war by the civilian population, despite many discouragements, is the subject of Becker's second book, *Les Français dans la Grande Guerre*. He finds that the French people's feelings about it disclose no perceptible pattern; certainly no trace of the constant keen dedication to the war that you would expect of any society one might wish to call thoroughly militarized. The qualified optimism of the early weeks of the war, he says, lasted into the end of 1916, and the French were more ordinarily patriotic than they were nationalistically exalted *à la mode de l'Union Sacré*. This mood lasted so long as there was still hope that the war might end fairly soon. It evaporated in late

1916, before the awful losses of 1917 gave Frenchmen a very good reason for wanting the war to end at the earliest possible moment. Becker notes that in 1917, France's worst year during the war, there appeared dissidents and a peace lobby, but not a positive wave of pacifist emotions nor any significant pattern in the antiwar agitations, such as they were. Even the revolutionaries in France apparently didn't want to achieve their revolution at the cost of or in connection with national defeat. Again we note a defensive patriotic rather than an aggressive vein of emotion. French persistence through 1918 he can only explain as a dogged holding on in a state of patriotic exhaustion from which, he surmises, the French had insufficient time to recover before the Second World War hit them twenty years later.

If militarization means a state of war-readiness in economy, mentality, and political and administrative organization, and if it means the ability to sustain that readiness through such a searching test as World War I put it to, the countries of Europe can be seen responding in different ways to the test of total war when against their expectations they got it between 1914 and 1918. Some of them cracked under the strain. The Russian Empire broke under a combination of internal strains and external discouragements. Austria-Hungary broke under similar strains. France did not break, though part of its army was close to cracking in 1917. Germany survived with its armed forces intact (although by that point they were losing the military struggle) until the very end, and with its civilian population fairly solid despite divers privations until waves of despair hit home as the military defeats became known on the home front and the extent of German withdrawals from occupied territory were realized.

The United Kingdom is the joker in this pack. Not the most militarist of the societies of Europe, one would have thought, when the war broke out, and yet by the end of the war, perhaps the most successfully militarized of the whole lot. Society, although divided and strained in rather the same ways that German society was, cohered better under the strain. The economy was successfully made over to a war effort, which by the last months of the war absorbed an enormous proportion of the people. And the huge armies that the United Kingdom put in to the field, although land forces had certainly not been considered Britain's forte before the war, became so "professional" and effective that during its last months they were winning big battles one after another as they pushed the Germans back toward their own country.

It would seem paradoxical that Britain, the society that best survived the strains of the First World War, was the nation that at the outset had every appearance of being the least militarized of the lot. But one must remember that militarization is not the same as militarism. Measured by the static ideal-type of militarism, Britain seemed wanting in many of the things necessary to sustain total war. But of its militarization there can be little doubt. However uneven and contradictory this process may have been, it produced a civil society successfully organized for the sustained violence of total war.

Chapter 2
Militarization and Rationalization in the United States, 1870–1914

PETER KARSTEN

AT CERTAIN MOMENTS in history the preparation for war comes to play a large part in the day-to-day life of society. Elements of civilian society come to adopt some of the values of the military. Stephen Wilson has recently speculated on these developments in fin-de-siècle Europe, developments that he calls "militarization."[1] In what follows, I will be attempting to answer the question of the extent to which the term "militarization" can also be used to describe developments in the United States during the same period.

There is no question but that the United States did experience some increase in "militarization" during the fin de siècle, but not as much of an increase as Europe experienced in the same period. Nor can we say that the United States was "militarized" at any time during these years. On one plane, it is clear that both the army and the navy of the United States grew both in size and effectiveness. Throughout these years the services streamlined their recruitment policies, articulated more fully their training institutions, created professional organizations and journals, and reformed their promotion, intelligence, procurement, and administrative systems.[2] The army's posts, 120 in number in 1888, were consolidated into 49 by 1912, allowing for training and maneuvers involving larger units. A Navy League (1903) and an Army League (1912) were created as lobbies. The National Guard grew as well, and in 1903 its federal appropriations were effectively tied to federal standards for the commissioning of officers, the recruitment and training of men, and the standardizing of equipment. Both the complement of and the annual expenditures for the U.S. regular army and navy quadrupled between roughly 1875 and 1910. This was not inconsequential, especially for the navy, which rose from being about the seventh rank among naval forces in 1880 to being the second, on a par with Germany, by 1908.

Moreover, the United States made increasing, systematic use of these augmented forces in its war with Spain and the Filipino insurgents, in its occupation of Cuba, in its acquisition of the Hawaiian Islands and the Canal Zone, and in its interventions in China, Nicaragua, Mexico, Haiti, and the Dominican Republic. These military expeditions were largely unplanned, but contingency plans for some of them had been drawn up, generally at the Naval War College or by the Navy General Board. The keel of a navy-industrial complex had been laid, and military reforms, particularly in the work of Secretary of the Navy B. F. Tracy and Secretary of War Elihu Root, Commanding Generals W. T. Sherman, John Schofield, and Leonard Wood, and Admirals Stephen B. Luce, George Dewey, W. S. Sims, and Bradley Fiske, had gone a long way toward transforming the American army and navy into a modern military machine.[3]

To what can we attribute this modernization and development of the American armed services, this rise in the extent to which the United States devoted its attention and resources to the preparation for war? Part of the answer is that there appears to have been an increase throughout the late nineteenth century in the extent to which civilian American society came to see military preparedness more favorably. Many of these "militaristic" civilians became active proponents of the modernization and augmentation of the armed services. The other, more significant part of the answer is that the opinion-making political elite in the press, Congress, and several administrations, most of them quite unmilitaristic, increasingly came to regard the modernization and augmentation of the armed services as necessary to secure and defend national interests, as they perceived and defined them.

The rise of this kind of unmilitaristic militarism presents something of a paradox and requires an explanation. Samuel P. Huntington and John P. Mallan have argued, to the contrary, that late-nineteenth-century America was characterized by a *lack* of militarism—that it was virtually pacifistic, dominated by an unmartial business mentality that was inhospitable to thoughts of war or military expenditures.[4] There is some truth to their claims, but they are greatly exaggerated. It is not only Theodore Roosevelt, Brooks Adams, and Homer Lea who offered a "warrior's critique of the business civilization." "The strenuous life" and "the martial virtues" were championed by many, among them: veterans of the Civil War, the Sons and Daughters of the American Revolution, organizers of the Boy Scouts, antiimperialists, authors of children's books, and persons who called themselves "pacifists"![5]

When one associates Civil War veterans and militarism, one must be prepared to respond to the query: "What about General Sherman? Didn't he call war 'hell'?" As Gerald Linderman has pointed out in *Embattled Courage,* so did many other Civil War soldiers, during and for some time after the war. Sherman's famous "war . . . is all hell" remark was made to five thousand Grand Army of the Republic veterans in Columbus, Ohio, in 1880. Linderman had no difficulty in finding numerous examples of such sentiments from officers and men between about 1864 and 1875; he found fewer such remarks with each year thereafter. As time passed, veterans mellowed; they repressed the memories of disease, boredom, ruthlessness, fear, death, and the struggle to survive. They restored their earlier vision of war as a test of manliness, a virtuous endeavor. So did Sherman. In 1890 he addressed veterans of the Army of the Tennessee in this way:

> No, my friends, there is nothing in life more beautiful than the soldier. A knight errant with steel casque, lance in hand, has always commanded the admiration of men and women. The modern soldier is his legitimate successor and you, my comrades, were not hirelings; you never were but knights errant transformed into modern soldiers, as good as they were or better. Now the truth is we fought the holiest fight ever fought on God's earth.[6]

The year these words were spoken, the GAR's membership peaked at 428,000, up from 30,000 in 1878, and 233,000 in 1884.

The most famous militaristic veteran of the Civil War is probably Oliver Wendell Holmes, Jr. His correspondence and speeches throughout the late nineteenth and early twentieth centuries are punctuated with bellicose phrases. Here, for example, are excerpts from his Memorial Day Address at Harvard in 1895:

> As long as man dwells upon the globe, his destiny is battle. The book for the army is a war-song, not a hospital-sketch, . . . War's . . . message is divine. . . . Some teacher of that kind we all need. In this snug, over-safe corner of the world we need it, . . . loving flesh-pots, and denying that anything is worthy of reverence. . . . I rejoice at every dangerous sport which I see pursued. The students at Heidelberg, with their sword-slashed faces inspire me with sincerest respect. I gaze with delight upon our poloplayers. If once in a while . . . a neck is broken [that] is a . . . price well paid for the breeding of a race fit for headship and command. . . . It is . . . necessary to learn the lesson afresh from perils newly sought. . . . to pray, not for comfort, but for combat.[7]

Holmes was not unique. The Civil War generation grew increasingly militaristic and offered an increasingly militaristic message to the next generation. One has only to open any of the published state or national proceedings of the conventions of the Grand Army of the Republic to see the remarks of delegates who could be mistaken for Holmes, calling for more military drill in schools, more flag reverence, more veneration of battle sites, and support for military expenditures to stimulate (in the words of one leader of the Society of the Army of the Potomac) "the increase throughout the land of a military spirit and of military organization and discipline." Moustaches, beards, or sideburns (the last allegedly named and modeled after the tonsorial choice of General Ambrose Burnside) became popular in the late nineteenth century. In 1887 the commander-in-chief of the Sons of Veterans society urged his members to "secure arms and equipments, and acquire as much proficiency as possible in the manual of arms and evolutions of the line." The circulation of the magazine *Century* doubled in 1888 when it began its "Battles and Leaders of the Civil War" series. In 1894 the commander-in-chief of the GAR encouraged youths to join volunteer military units and urged his veteran colleagues to do the same:

> Visit the armories of our National Guards, encourage the best class of young men to join the ranks, invite them to our camp-fires and our Memorial Day services. Assure them that the soldiers that were are in full sympathy with the soldiers that are, and will support them in the discharge of their duties to the fullest extent.

Senator John Thurston of Nebraska, the son of a veteran, told a Lincoln Day dinner audience in February 1895:

> Let the true story of every American battlefield be taught in every public school. Set the stars of the Union in the hearts of our children, and the glory of the republic will remain forever. It does not matter whether the American cradle is rocked to the music of "Yankee Doodle" or the lullaby of "Dixie," if the flag of the nation is displayed above it, and the American baby can be safely trusted to pull about the floor the rusty scabbard and the battered canteen, whether the inheritance be from blue or gray, if from the breast of a true mother and the lips of a brave father, its little soul is filled with the glory of the American constellation.[8]

Militarism of this sort, brimming with nationalist sentiment, veneration of Civil War service, and an admiration for the martial virtues could be found in many corners of American culture in the late nineteenth century. We are not surprised when we find it in the pages of the *U.S. Naval Institute Proceedings*, the *Infantry Journal*, the *Army and Navy Journal*, or *Seven Seas* (the Navy League's publication). Readers of these journals were told that "the true militarist believes that pacifism is the masculine and humanitarianism is the feminine manifestation of national degeneracy," and that "every boy . . . is eager to go out and beat his neighbor in . . . some . . . game, representing the ever-continuing conflict of earthly existence" and "does not care about being told that a cannon shot costs more than a small house . . . and more of that kind of misleading talk." They were told that "military discipline is the most potent influence around because it is associated with those martial yearnings which every properly constituted boy entertains at some stage of his life. . . . The drum and fife and glistening gun to such a boy mark the highest forms of achievement." Readers of "preparedness" *defensoribi* like Hudson Maxim's *Defenseless America* (1915) learned that "German militarism is the greatest school of economics that the world has ever seen," and that "conscription . . . makes good citizens . . . and creates a greater respect in them for ruling power and for law and order," and that military drill was a "manly game."[9] But one would expect that the speeches, journals, and books produced by the armed services and their allies in those years would say such things. After all, most of these officers clearly favored the militarization of America and were decidedly wedded to antimaterialistic, martial values and Darwinian metaphors (something I have called "national Darwinism.")[10] For example, in the libraries of the "Great White Fleet" in 1908 one found the works of Mahan, Herbert Spencer, Rudyard Kipling, and John Fiske well represented.[11] Some naval officers spoke of "man" as "a fighting animal, carnivorous and therefore with killing propensities. . . ." Others spoke of war as "the acme of the endeavor of man," a force that "quickened the pulse of the nation" and "sent a brighter, stronger current to eliminate morbid germs from all the tissues of the body politic, offsetting tendencies toward commercialism and materialism." "You can be dismissed neither by politics nor business conditions," one rear admiral told the midshipmen during a depression, "World disarmament alone can affect you." Some officers admired "the submission" and authoritarianism of "the Prussians"; others, Napoleon and those of "unrelenting heart who tear away and break down every obstacle." And most identified with Rear Admiral Edward Simpson, who spoke in 1888 of the navy as the institution "in which has been merged my whole identity, *for* which I have worked, and *in* which I take my greatest pride, . . . the mistress I tried to adorn, whose favors alone I coveted, for whom I could never do too much. . . ."[12]

We are not surprised by militarism within the American officer corps or within its lobbies, the Army and Navy Leagues. We should not even be surprised that a "cult of Napoleon" flourished in these years, or that there were over one hundred thousand men in the entirely voluntary National Guard units by 1895 and some twenty-nine thousand volunteers in the army's military programs at the land grant colleges by 1911. In 1885 the Odd Fellows (a fraternal and benevolent society) created an auxiliary, "the Patriarchs Militant," and it is noteworthy that by 1897 there were over two hundred fifty thousand members of that organization, parading in splendidly ornate uniforms.[13] But how much are we to make of this? After all, there have always been some in every society who enjoy uniforms, drill, and things military.

What is more noteworthy is the presence of militarism or the admiration of military values among antiimperialists and self-styled pacifists. Let us consider the famous essay of one such figure, William James, "The Moral Equivalent of War" (1910). This is the essay that is said to have inspired the American Friends Service Committee, the Civilian Conservation Corps, the Peace Corps, and Vista, and served as the motto for President Jimmy Carter's energy conservation program. It certainly aims at war avoidance ("The martial type of character can be bred without war," James writes.) But its premises bear noting. James believed that mankind is inherently warlike:

> Our ancestors have bred pugnacity into our bone and marrow, and thousands of years of peace won't breed it out of us. The popular imagination fairly fattens on the thought of wars. . . . Militarism is the great preserver of our ideals of hardihood. . . . There is a type of military character which everyone feels that the race should never cease to breed, for every one is sensitive to its superiority. . . . Martial virtues must be the enduring cement.

Notwithstanding James's objectives, this essay evinces his acceptance of particular models of human nature and of society that were decidedly militaristic, around which James felt that he had to work in order to create his "moral equivalent of war." Moreover, there is corroborating evidence of James's acceptance of militarism; we need not restrict ourselves to this essay. In his correspondence we find occasional reference to "the old fighting instinct that lies very near the surface in all of us." The Spanish-American War, he observed, was to Americans "a peculiarly exciting kind of *sport*."[14] If William James, an antiimperialist and professed antimilitarist, believed these things, or even if he was simply convinced that his countrymen believed them, it is possible that these values did pervade society.

Edward Bellamy's best-selling neosocialist novel, *Looking Backward* (1887), may serve as another example of evidence of militarism hidden in what may seem to some an unlikely place. Bellamy's hero, Julian West, lives in Boston in the year 1887. He is witness to a military parade, in which he sees "order and reason." West awakens from a mysterious sleep one hundred twelve years later, in the year 2000, to discover a utopia in which "honor," bestowed by society on the good worker or virtuous citizen, was "the sole reward of achievement, thus imparting to all service," the hero observes, "the distinction peculiar in my day to the soldiers." Bellamy, a minister's son, who would be a Populist in the 1890s, had unsuccessfully sought entry into West Point as a youth. He studied for two years in Germany in the 1870s and came away with great admiration for the German military and its social dimensions. Later he would praise "the military systems of the great European States, . . . wonderful examples of what method and order may accomplish in the concentration and direction of national forces." Bellamy, a man who had never seen war, editorialized about war in his short story, "An Echo of Antietam," in 1889: "What a pity it is that the tonic air of the battlefields . . . cannot be gathered up and preserved as a precious elixir to reinvigorate the atmosphere in times of peace, when men grow faint and cowardly and quake at the thought of death." He possessed "an admiration of the soldier's business as the only one in which, from the start, men throw away the purse and reject every sordid standard of merit and achievement."[15] In his novel, the utopian world of the future had a single political party, the

Nationalist Party, which held military musters of its members. It would be stretching things to call Bellamy a militarist and a proto-fascist, but not stretching them very much.

There is also abundant evidence of unmilitaristic militarism in the newspapers of the day. The press was no unitary actor; one can find many different views expressed in the editorials of America's newspapers of the period. And it is not clear that there is a strong direct correlation between public opinion and editorial views. Obviously, editors worry about circulation and are wary of presenting views that are often out of synch with their readers' opinions, but for reasons we cannot enumerate here, editorials cannot be presumed to be a clear guide to popular views. In the absence of public opinion polls, however, editorials, letters to the editor, and reporters' views are of some value. Once again, one can find antimilitarist and antiimperialist sentiments in major papers in the *fin de siècle.* But one can also find blood-curdling language, language stronger than anything offered by today's most jingoistic journals. The "yellow press" on the eve of the Spanish-American War is infamous. The congressman who maintained in early 1898 that each of his colleagues "had two or three newspapers in his district—most of them printed in red ink . . . shouting for blood" was not exaggerating very much. The *Washington Post* editorialized regarding the war atmosphere:

> A new consciousness seems to have come upon us—the consciousness of strength, and with it a new appetite, a yearning to show our strength. . . . ambition, interest, land-hunger, pride, the mere joy of fighting, whatever it may be, we are animated by a new sensation, . . . the taste of blood in the jungle.[16]

In short, a reading of fin-de-siècle American newspapers does provide further evidence of a militaristic, Darwinistic spirit.

How important is this spirit of militarism, discernible in the officer corps, the volunteer units, the Civil War veterans' organizations, children's literature, the press, and other nooks and crannies of fin-de-siècle America?[17] Stephen Wilson believes that "the existence of military values in society-at-large in particular," constitutes "the necessary condition at least of the preparation for, and the practice of, mass warfare." The militarization of civilian society, he maintains, "is clearly a crucial factor in explaining why large-scale wars have and can be fought. . . ."[18] This sounds right, but it may overstate things just a bit. Civilian militarism is a necessary condition for massive war preparations and for mass warfare, to be sure, and it is a factor in explaining such phenomena as America's declaration of war with Spain, early twentieth-century military interventions, and American intervention in World War I. But Wilson gives civilian militarism more weight than I would. It is only one factor (and usually not a crucial one) in explaining why wars are fought, and it is only a necessary condition to militarization and mass warfare, not a sufficient condition (which I take to be what Wilson is hinting at when he calls it "at least" a necessary condition.) More important, in my judgment, than civilian militarism is the more rational, cool, deliberate vision policymakers hold of the national interests in the decision to embark on extensive military preparations or mass warfare.

In the case of the United States in the fin de siècle, it was not the civilian militarists any more than it was the uniformed militarists who commited the nation to programs of military modernization and growth or who plunged the nation into war. Assistant Navy Secretary Theodore Roosevelt did not produce the victory at Manila Bay by ordering Commodore Dewey to Hong Kong for coal while his less bellicose boss was out of town. He did these things, to be sure, but as John Grenville, George Young, and Ronald Spector have pointed out, Secretary John D. Long was justifiably annoyed with T. R. when he discovered what his subordinate had been up to. "He's gone at things like a bull in a china shop," Long wrote. The Naval War College had already drawn up plans for an attack on the Spanish fleet at Manila Bay; arrangements had been made for coal; Roosevelt's frenzied messages were only going to muddy things up.[19] William McKinley and the fifty-fifth Congress certainly felt public pressure to aid the Cuban insurgents and to "give Spain a thrashing,"[20] but, as Walter LaFeber has demonstrated, McKinley resisted that pressure until he was persuaded by the course of events, by the reasoned judgment of influential national and financial leaders (such as the *Journal of Commerce,* and the *Wall Street Journal*), and by his own calculations that the vital interests of the United States were being unjustifiably imperiled by the continuation of the Spanish-Cuban War.[21] He had available to him a suitable navy in part because of the navalist zeal of the officer corps and its allies. But in larger measure McKinley and his predecessors possessed an ample navy because these navalists had a clear sense of *Realpolitik*—that is, they presented a case for naval modernization, growth, and "war readiness"[22] that many in the Congress, the administration, and elite opinion-makers could appreciate, one that was keyed to national interests as they were understood by those elites (interests like the China market, coastal defense, the movement for a ready navy, and Caribbean hegemony), and one that was, after all, moderated and cut down to size by those elites.

Army modernization also unfolded to the extent that its advocates offered persuasive arguments to those same political and economic elites. General John Schofield, for example, spoke not of glory or manliness or the purgative virtues of war as commanding general of the army or, earlier, as commander of the Division of the Atlantic. He spoke rather of "the wise scientific use of surplus reserve." He noted that "the sacrifice of the brave," "the valor of great masses of men, and even the genius of great commanders in the field" had been "compelled to yield the first place in importance to the scientific skill and wisdom in finance which are able and willing to prepare in advance the most powerful engines of war."[23]

The same forces that rationalized and modernized the armed services also rationalized and modernized the foreign service[24] and the nation's planned use of its natural resources,[25] organized a rational, "informal" empire,[26] participated in a "rational" arbitration movement,[27] and eventually, for the same reasons and in the same fashion, produced a military draft. One of the conscription movements' most vigorous opponents, Amos Pinchot, described in perjorative and only partially accurate language its objectives in March 1917:

> Conscription is a great commercial policy; a carefully devised weapon that the exploiters are forging for their own protection at home, and in the interests of American financial imperialism abroad. . . . Back of the cry that America must have compulsory service or perish, is a clearly thought-out and heavily backed project to mould the United States into an efficient, orderly nation, economically and politically controlled by those who know what is good for the people.[28]

J. Garry Clifford, John Chambers, John Finnegan, and Robert Cuff have demonstrated that the "preparedness movement" on the eve of World War I, the conscription movement, and the management of the economy during World War I were championed by rationalistic, cosmopolitan elites with a pragmatic vision of the nation's defense needs, its social, ideological, and economic well-being, its resources, and its antimilitary traditions.[29] We may not agree with that vision; we may not see the same risks to the nation's interests that those elites thought they saw, but we must allow that few if any of them were raving militarists. For example, the founders of the National Security League formed in December of 1914 were, in the words of the *New York Times,* "sober-minded . . . merchants, financiers, professional men of all shades of political opinion." Preparedness was to them a necessary evil. "This is simply a business proposition," a League official explained. Corporate securities, financial transactions, America's share of world trade, indeed, American prosperity would suffer in the same fashion as had the warring nations if the United States were threatened or attacked by the victors of the world war.[30] To the extent that this view was shared by elites in the United States—and I believe that the evidence demonstrates that it was widely shared by those elites—the "preparedness" movement of 1910–1916 as well as the "preparedness" efforts in the fin de siècle were the sober products of relatively unmilitaristic minds, the minds of men who preferred arbitration to war, civilian values to military values, adequate military force to "superiority" and the goosestep.

That this is so is demonstrable. It is true that some elites longed to see the United States enter the imperial race; some craved the status of a "world power" largely for its own sake. But most did not. Yes, the United States created the rank of "ambassador" in 1893 to improve the status of its "ministers" to foreign nations. Yes, it created more rear admirals and brigadier generals at the turn of the century to improve the status of its military representation. But it did not do so because it was bent on imperialism or war. It did so to elevate the dignity of the United States at official receptions and ceremonies, because such status had practical and demonstrable consequences for American diplomats and commanders in specific situations. Yes, U.S. military personnel were increased in the fin de siècle, but more in absolute terms than as a percentage of the United States labor force. While the numbers of active duty military personnel rose from fifty thousand in 1870 to one-hundred-thirty-nine thousand in 1910, this was a rise from .03 percent to .04 percent of the entire labor force.[31] The United States spent several times as much in real dollars on its warships in 1915 as it had in 1875, but its naval status actually declined in the decade between 1905 (when HMS *Dreadnought's* keel was laid) and 1914.[32] In that first decade of superbattleships, the United States slipped from second to a poor third behind Germany and Britain, with eight of the new battleships and battle cruisers compared with twenty-one for Germany and thirty-four for Britain.[33] If there was a naval arms race in this decade, the United States was not in it. And this is so despite the fact that the industrial capacity of the United States measured in terms of its iron and steel output had increased at a rate twice as fast as that of Germany between 1875 and 1914, over six times as fast as that of France, and almost fourteen times as fast at that of Britain. Moreover, the iron and steel output of Britain, France, and Germany combined in 1914 exceeded that of the United States alone by only a tiny fraction (64.3 million tons versus 62 million).[34] By the early twentieth century, the United States had the manpower, the wealth, and the industrial capacity to lead all others in military might. It lacked the disposition to do so. Many of the "preparedness" advocates favored universal military service, some for reasons that can be styled "militaristic," but only Germany's declaration of unrestricted submarine warfare in early 1917 and the vigorous efforts of President Wilson and his conscriptionist allies induced Congress to reverse its position and adopt the draft, a full month and a half after it declared war.[35] Yes, thousands of young men joined drill companies or the Patriarchs Militant or land grant college military units or summer military training camps, but evidence of this sort of civilian militarism is not limited to the fin de siècle. As Marcus Cunliffe has demonstrated, such fascination with things military was quite common in antebellum America, in the North as well as the South.[36]

In short, by any standard the United States was less militarized than Europe in the fin de siècle. Its culture was less militaristic and its government devoted fewer resources to preparations for war. This was so, in part, because the United States possessed no feudal past; in part, because its inhabitants retained, well into the twentieth century, a fear of standing armies. More significantly, it was so because, unlike Germany, Austria-Hungary, France, Italy, or Russia, it did not have a common border with a potentially dangerous land power.[37] Until the 1940s we enjoyed three thousand miles of what C. Vann Woodward has styled "free security." Carl Schurz put it well. In 1893 he wrote in *Harper's Magazine* that

> No European enemy could invade our soil without bringing from a great distance a strong land force; and no force that could possibly be brought from such a distance, were it ever so well prepared, could hope to strike a crippling blow . . . without the certainty of being soon overwhelmed by an easy concentration of immensely superior numbers. . . . In other words, in our compact continental stronghold we are substantially unassailable.[38]

This allowed the United States to avoid the militarization that gripped Europe for decades before 1914. This is what made

Hudson Maxim's preparedness manual, *Defenseless America,* (1915), seem less than persuasive; it is this that made the 1916 film, *The Battle Cry of Peace,* based on Maxim's book depicting spike-helmeted invaders sweeping across the New Jersey countryside, seem implausible to others. As Maxim himself wrote: "The people are imbued with the belief that the country as a whole is big enough and prosperous enough to be safe." Maxim disagreed with that judgment, but he acknowledged it. He also acknowledged that "militarism" would never sell: "I am not arguing for a large standing army, but merely for an adequate army [which he defined as being 200,000 to 250,000 strong]."[39] Army Chief of Staff General Leonard Wood took the same view:

> We do not want to establish militarism in this country. . . . But we do want to build up in every boy a sense that he is an integral part of the nation, and that he has a military as well as a civic responsibility. All this can be done without creating a spirit of militarism or of aggressiveness.[40]

Our policymakers did modernize and augment our naval and military forces in the fin de siècle, but they did so, or were forced to do so, with some moderation. America's moment of militarization would come, but not for another half-century. It would come when the United States and the Soviet Union faced one another as major powers in a world grown more hostile due to ideological differences, smaller due to intercontinental bombers (and, eventually, intercontinental missiles), and more deadly due to nuclear weapons. "Rational," "pragmatic" civilians, not "militarists," would be the progenitors of America's midcentury militarization.

PART TWO
Militarization and the World Wars

Chapter 3
Toward a Warfare State: Militarization in America during the Period of the World Wars

PAUL KOISTINEN

ATTEMPTING TO ANALYZE militarization, defined as the impact of war preparation on society, during the period of the Great Wars is a formidable, daunting task. Besides the magnitude of the assignment, other significant problems exist. The foreign policy that shapes military policy, for example, can only be treated in passing. Also, reliable data on public opinion is not always available nor has it been properly analyzed. Moreover, the press of time and space requires the use of simplified labels, such as "Neo-Jeffersonian" and "Neo-Hamiltonian." Additionally, many generalizations will seem unsatisfactory without detailed substantiation. Finally, since this is a subject that has seldom been treated directly, some, at times much, of what can be said may be considered as suggestive rather than definitive.

This discussion will be divided into four parts. The first deals with the "preparedness" movement from 1914 to 1917. Here we see the nation attempting to deal in a very confused manner for the first time with the reality of modern warfare. In the second section, the war years will be examined. The major problems encountered were those involving military command and economic mobilization. The interwar years, as the third part, are the most significant in a sense. On the surface, preparations for war seemed slight. Yet, very significant adjustments to war were taking place in terms of the armed services command systems and in terms of strategic and economic planning for future wars. Also, a far-reaching and very astute critique of modern warfare that was begun during World War I continued into the 1920s and 1930s, culminating in the work of the Senate Special Committee Investigating the Munitions Industry (called the Nye Committee). In an analysis of World War II, as the final part, I demonstrate that the years from 1914 to 1940 had been very significant for war preparedness. By comparison with the First World War, the Second was conducted much more effectively at home and abroad. Furthermore, for the American public this war was much less traumatic than the earlier one.

Throughout these four stages, several themes recur, and these need to be introduced now. Modern warfare involves basic issues of political economy. Hence, it had a significant effect upon the reform movements of the time, which were seeking to define the proper role for the government as economic regulator, social welfare agent, and the like. Both war and reform also involved critical issues of power in American society. The conduct and preparation for war became increasingly elitest in this period in that they were directed or shaped by the federal executive and by large corporate structures working with the armed services. This led some reformers and spokespersons for the masses to insist that modern warfare was undermining both the reform and democratic traditions. Certainly, war and preparation for it acted to strengthen nationalistic and patriotic forces, which became very abusive during and after both world wars.[1]

The movement for war preparedness from 1914 to 1917 was curious and paradoxical. Men like Henry L. Stimson, General Leonard Wood, and Theodore Roosevelt wanted an enlarged military either because it was the logical need of a major power or in order to intervene in the European war. Another and much larger group of preparedness enthusiasts seemed to favor the cause as much or more for purposes of social control at home than for purposes of war abroad. These were the urban, professional, and business Neo-Hamiltonian reformers who were interested in disciplining if not regimenting American society, homogenizing the immigrant masses and rationalizing the political and economic order. Their efforts became concentrated in organizations such as the National Security League and the Plattsburg Movement. Most ultimately became backers of universal military training (umt) and National Service Legislation as well.[2]

The preparedness drive and the outbreak of war in Europe helped to create the modern peace movement, which was beginning to divide along the lines of the pacifists versus the internationalists. The former, of course, opposed the war, while the latter, which grew out of such establishment organizations as the Carnegie Endowment for International Peace, ultimately supported the war and often did so avidly.[3]

But most of those opposed to preparedness did not join peace societies. The biggest block of opponents came from rural areas, and they were either hostile or indifferent to an enlarged military. Organized labor stood quite solidly with the farm community. Antipreparedness, therefore, involved primarily rural America and the urban lower- and lower-middle class elements. Urban middle- and upper-class antipreparedness sentiment was usually associated with social welfare (as opposed to efficiency) advocates and with Progressives, along with radicals and pacifists.

President Wilson, who headed an administration divided on preparedness matters, at first opposed an expanded military and professed neutrality until just before the nation entered the war; but he sought a middle way on preparedness after the *Lusitania* crisis in May 1915. With the National Defense Act of 1916, the army was to be expanded modestly and gradually, and the American navy made second to none and probably ahead of all—an entirely new departure in American strategic policies.

Wilson's acts were as much political as military. In significant ways, diplomacy and strategy were uncoordinated. The general staff was planning for a transoceanic invasion by Germany, Japan, or other hostile powers, and the navy's plans for expanding its fleet and fighting a war were equally unrealistic. These dissociations were characteristic of the entire debate over preparedness. While the opponents made direct, forceful, and candid arguments against military expansion, the proponents, because of the political and military realities they faced, often argued their case indirectly, by making appeals to patriotism, even jingoism, and by attempting to instill fear or panic in the populace.

What the nation needed was a new body for coordinating its

foreign and military policies and for handling national security issues that were exclusively neither civil nor military. A defense or war council, composed of members from Congress, from the executive branch, and from the armed forces had been proposed for that purpose in various forms since the late nineteenth century, but without any real chance of succeeding. Such a body might have begun to focus the preparedness debate.[4] In retrospect, such an agency was not adopted, less because of military policy than because of foreign affairs. From the late nineteenth century forward, American diplomacy has been either so uncertain and contradictory or, in terms of the position of the leadership toward the led, so duplicitous, that obscurity, not clarity, was necessary or preferred.

Despite its odd and perverse qualities, the World War I preparedness movement did help to ready the nation for hostilities. Over two years of incessant advocacy for a big military had begun to condition the people for total war. More concretely, the agitation for a continental army and universal military training led the general staff to draft plans for training millions of men in a relatively short period of time. And the Plattsburg experiment provided the basis for the large number of reserve officers that the new army would need. The army, therefore, was ready to implement the Selective Service System. That system was much less harsh than universal military training. Furthermore, the use of local draft boards made Selective Service more palatable to the public and disguised the centralized military control that was involved. As it turned out, raising and training troops were not the major obstacles to efficient mobilization. Commanding, transporting, and equipping the army were the most trying tasks.

In 1917 both the army and the navy had relatively new command structures. The army's chief of staff–general staff system had never worked as it was intended before 1917 and it did not take hold during the war until General Peyton C. March in 1918 rebuilt the office he headed into a true command center. Consequently, Secretary of War Newton D. Baker exercised exceptional control over the fighting forces in 1917. Yet his reach was never far enough. General John J. Pershing created his own command system in Europe, which had more power and prestige in traditional military terms than the one that existed in Washington. Had the war stretched out, it is questionable whether Baker would have backed March in reestablishing Washington's control over Pershing, a situation that could have become critical. The Navy Department's problems were less pressing because it was smaller in size, had expanded less, and had a more modest role in the war. Moreover, Secretary of the Navy Josephus Daniels worked well with his chief of naval operations and supported and assisted by Wilson he managed to maintain reasonable control of naval policy in Washington.

A short war prevented the issues involving civilian control over a modern military from escalating to the point of crisis. However, the crucial matter of military demand for munitions and supplies was never satisfactorily resolved, despite the fact that these factors were basic to any mobilization system. Military commanders did not want civilians to review their requirements and the civilian mobilization agencies were loath to get into that area. Here was a perfect example of where a war council was needed to settle a critical issue requiring both civilian and military expertise.

The largely unanticipated, and during the preparedness debate, relatively uncontroversial problems of World War I involved economic mobilization. Despite the experience of the European war, few saw clearly that an effective wartime planning mechanism was essential. For the United States, this planning would combine economic and military elements on a major scale in a way that would have crucial consequences for the future. Some of the problems had at least been suggested with the building of a modern navy. That accomplishment had required putting together a new production team involving the executive departments, the military, and industry. In the late nineteenth and early twentieth centuries, however, the new navy's limited scale failed to reveal the significant problems that the political economy of warfare would raise in modern times.

The economic mobilization process began inconspicuously in

mid-1915, took on serious proportions when the nation entered the war, and reached maturity with the War Industries Board (WIB) early in 1918. By that time all of the elaborate controls necessary for harnessing the economy had been introduced. With the armistice, the system was not fully integrated and coherent but the basic structures and methods had been developed and used and could be refined.

Economic mobilization created the most intense civil-military dispute during the First World War. As the major claimant agencies, the armed services, out of parochialism, weak leadership, and fear of losing their prerogatives, resisted adjusting their systems to those of the civilian mobilization agencies, which in turn were patterned after the basic structure of the economy. The army was especially culpable in this regard and its actions threatened to create economic chaos. A desperate corporate community prevailed upon Congress to pressure the Wilson administration into working out a compromise during the winter of 1917–18. The military kept control of its procurement systems, but patterned them after the WIB and integrated them into that agency. This solution allowed the American military to exercise greater control over its own procurement than most other belligerents considered expedient.

The decision to allow the military to continue performing its own procurement grew out of the nation's political economy. Agitation over consolidated economic power and abusive corporate capitalism was still strong when America entered the fray. The methods used to mobilize the economy intensified this debate. Businessmen serving both as private and public representatives dominated the various mobilization agencies to the point of indirectly awarding contracts to themselves. Few safeguards existed to protect public interests. Neo-Jeffersonians questioned the system, but they were held off by the plea of military necessity. Removing procurement from the military would have required an entirely new mobilization system; otherwise industry would have been awarding public contracts directly to itself. Any untried or different system, however, held little promise of success. As long as the armed services procured legally, this helped to give the appearance of private and public interests being separate. Critical matters of political economy were thereby avoided.

Another World War I trend is significant: military policy and ideology were combining to create volatile and unpredictable political forces. War and revolution had become practically one in the twentieth century.[5] That reality eventually brought out the intolerant, absolutist underside of Progressivism in which all of the tensions, ambiguities, and paradoxes involving war and the military were projected into a crusading, abusive nationalism and patriotism. The hysteria was abetted at the end of World War I by the Wilson administration's initiation of a naval race that threatened its allies and its own people with the choice between a new world order or an armed and isolated America.

By standard measures, the impact of the military on society in the interwar years was not great. In unadjusted dollars, appropriations were larger than before the war, rising from 245 million dollars in 1913 to 1,567 million dollars in 1940. But in terms of percentages of total government spending, military expenditure fell from 28.3 percent to 15.5 percent in the same period. A consistent pattern of expanding military budgets began only in the late 1930s, and even in 1940 the amount was not exceptional.[6] Moreover, the armed services appeared to be looking backward, not forward. For example, they did not move aggressively to adopt and adapt to the new technologies of warfare. The army did little with the tank and other motorized weapons and transport, with advances in artillery, small arms, and so forth, and the navy appeared similarly hidebound. Only in air power were the services progressive, especially the navy. Army progress was probably retarded by the drive for an independent air arm. Throughout the interwar years, military demand was fundamental to most aircraft and aeronautical developments. Nonetheless, the major breakthroughs came from civilians, backed by the government through air mail contracts, regulation, and support for airports and other facilities and services.[7]

Despite limited spending and technological advancement, great strides were made by the military, especially the army, during the 1920s and 1930s in preparing for a modern war. Chief of Staff General Douglas MacArthur, 1930–35, and his successor, General Malin Craig, 1935–39, used the National Defense Act of 1920 to prepare the land forces for fighting another major war not based on the traditional military devotion to large standing armies, but based instead on citizen soldiers, raised by enlistment, by the reserves, the National Guard, and selective service. Additionally, after years of being totally unrealistic, the general mobilization plans for raising massive armies were finally based on the economy's potential. Consequently in 1940 the army was better prepared for hostilities than had ever previously been the case. During these years, the navy, always in a state of greater preparedness, concentrated on strategic planning. Late in the 1930s, the army was ready to join the navy in this regard, and in 1939 the Joint Army and Navy Board wrote a series of so-called Rainbow Plans, the last of which set forth quite accurately the American strategy for fighting World War II. The military planning in the interwar years stands out in stark contrast with that which occurred before the First World War.[8]

An efficient command system made this planning possible. With the National Defense Act of 1920, Congress took a more positive attitude toward the general staff than it had taken in the past. That permitted General Pershing, chief of staff from 1921 to 1924, to fashion a new staff system, including a War Planning Division, which was responsive in peacetime and could be and was made into a command center in Washington, not abroad, for conducting a global war. The Navy Department also made some progress along these lines. Together the services' Joint Planning Board provided the basis for the World War II Joint Chiefs of Staff.

Perhaps the most critical aspect of the National Defense Act of 1920 was the creation of the Office of the Assistant Secretary of

War (OASW) to plan for procurement and economic mobilization in the event of another war. In that way the nearly disastrous operations of the War Department during the First World War could be avoided. By slow stages from 1921 through 1929, the OASW guided the process of having all the supply subdivisions of the army study and write plans for wartime procurement. It then took up planning for the entire economic system in what became four increasingly sophisticated industrial mobilization plans written between 1930 and 1939.

This planning was of the utmost importance for several reasons. First, it was basic to realistic mobilization planning because it forced the strategic planners to accept their dependence upon harnessed economic might. Second, it brought the army into contact with the business-industrial community in an ongoing, extensive, and often intimate way. The relationship ranged from the largest, oligopolistic industries such as steel to the miniscule optical glass business. Additionally, practically every trade association of any significance was at least in contact with the OASW and many were deeply involved in the planning. The same was true with all major business federations, such as the Chamber of Commerce.

Finally, the OASW planning resulted from and nurtured a new type of organization intended specifically to help bridge the gap between the civilian and military worlds in the area of munitions. The best example of this is the Army Ordnance Association (AOA), organized in 1919 by some of the same individuals responsible for creating the OASW, the most important of whom was Assistant Secretary of War Benedict Crowell. The association played a key role in getting the procurement and economic planning under way, and it consistently acted or attempted to influence War Department policies and practices. The Navy League, organized in 1902, was a prototype for the AOA, but the latter, including major munitions firms, related industries, and active and retired army officers, was much more dynamic and influential. On a broader scale, more prestigious and even more involved was the National Advisory Committee for Aeronautics, created in 1915. These organizations set a pattern for the plethora of such groupings that exist today.

Their motivations, like those of industry in general, ran the spectrum from dedicated public service to unmitigated greed.

A mounting critique of the political economy of warfare grew along with military preparedness in the 1920s and 1930s. The agitation actually began as early as the late nineteenth century with the building of a modern navy, and it took on new life and urgency during World War I. Emergency conditions always limited what was done. After the war, however, the critical scrutiny began and it never actually ceased until America's entry into the Second World War.

The last major investigation was run by the Nye Committee from 1934 to 1936. The committee's work presents a representative picture of the entire wartime and interwar critique. While Nye and his colleagues covered much of the same ground as their predecessors had, they also went beyond them in documents and practices reviewed, analyses made, and conclusions drawn. In view of the restraints under which it worked, its limited funding and staff, and the hostile and powerful enemies it had, the accomplishments of the Nye Committee were extraordinary. Contrary to conventional wisdom, Nye and his colleagues and staff made a major contribution toward understanding the impact of modern warfare on American society.

One of the committee's most important findings was that modern warfare and preparation for it had wiped out the lines of demarcation between civil and military, private and public institutions. When focusing on munitions alone, the committee found all of the institutions and practices, albeit often in microcosm, that a later generation would call the military-industrial complex. Industries and the military worked for their mutual benefit and often at the public's expense; they exchanged personnel and they struck bargains on profits, pricing, and monopolies. These interests, joined at times by the State and Commerce Departments, consistently worked to undermine national policies for arms control, arms sales, and arms embargoes. Members of what the committee called an "unhealthy alliance" looked to defense spending to help cushion the effects of the depression, and federal funds were used for that purpose. In examining the OASW's industrial mobilization plans, the Nye Committee argued that the planning only assured that industry and the military would work together more effectively than had occurred during the First World War. That meant the conflicts of interest, gross profiteering, waste, inflation, and other political and economic "evils" would probably be even greater than before. When focusing on the neutrality period of World War I and the operations of J. P. Morgan & Company as purchasing agent and financier for the allies, the committee revealed for the first time how deeply involved the house of Morgan had become in the decision making of Washington, London, and Paris, which helped make a mockery of American neutrality. When Senator Nye drew the logical and indisputable conclusion that Wilson and other top government officials had lied to Congress about relations with the allies, he opened the way for detractors to shut down the committee under the guise of respecting the dead.

Several points about the Nye Committee need to be made. First and foremost, it represented primarily the rural critique of preparedness and war that had begun decades ago. The Neo-Jeffersonians were protesting not just against war but also its effects on society. Warfare had become the ally of the Neo-Hamiltonians. During the First World War, economic planning went beyond the anticipation of its most avid proponents, and it set the pattern for the informal planning of the Republican administrations in the 1920s and the direct planning of New Deal agencies in the 1930s. These results did not move America closer to the good society anticipated by some reformers and radicals. Instead, planning tended toward regimentation and social controls such as the "work-or-fight" measures of World War I. The same tendencies were evident in the 1920s, when during a period of intense antilabor sentiment, the American Legion and others cooked up proposals for drafting labor in future wars, justifying these plans under the banner of "universal service," "equalizing the burdens of war," and "taking the profits out of war." Finally, according to the Neo-Jeffersonians, war had further concentrated economic power and had so dislocated the nation's economic system that it was a major cause of the Great Depression. All of this meant that warfare had and would have a greater impact on twentieth-century America than reform would have unless drastic steps were taken to change direction. The Nye Committee analysis was of such a nature that it attracted and won the support of a large segment of the American population, including urban liberal reformers, radicals, peace advocates, pacifists, isolationists, and those of like thinking.[9]

Second, the Nye Committee's close scrutiny of the OASW's plans and planning revealed that the military was hesitant and suspicious about working with industrial America. It feared that the commercial ethic, the profit motive, and greed would undermine the military mission and its professionalism. Yet, the planners felt that they had few alternatives. Any mobilization scheme would be dominated by corporate America and the armed services had to go along in order to get what they needed from the system.

Third, the Nye Committee became convinced that the only way to avoid all of the baneful consequences of warfare was to avoid war and excessive preparations for it as well. Finally, the starkest truth about the Nye Committee was that all of its efforts, analyses, and revelations appeared meaningless in the face of the larger, seemingly inexorable forces of war and the preparation for it. By the time the committee had concluded its work, the Roosevelt administration had begun its naval expansion and was only a few years away from beginning to mobilize the economy for another war.

By using World War I methods for harnessing the World War II economy, the administration revealed the superficial nature of the New Deal as a reform movement. The New Deal had not altered the American political economy in any significant way. Moreover, the war served to restore industry's depression-tarnished image.

During the Second World War, the strife over economic mobilization was much greater than it had been during the earlier war, but it was different in nature. Basically, the conflict involved interest groups and class structures, rather than civilians and the military. By slow stages between 1939 and mid-1943, the nation's largest corporations and the armed services devised a working alliance that served their mutual needs and that was institutionalized in the War Production Board. Corporate America got a commitment to the system as it operated, and the fighting forces received, within reason, unquestioned support for their requirements. This system, protected by southern Democrats like James F. Byrnes holding key governmental positions, operated against the interests and power of labor, small business, the consumer, and the so-called New Dealers who favored more efficient, more balanced, and more equitable mobilization policies. The population, of course, did not suffer excessively; in fact it experienced a rise in income, savings, and consumption. Nonetheless, the programs that were devised to assist and facilitate the masses migrating to war production centers were minimal; inflation was controlled by holding a lid on wages and profiteering was visibly high. These realities generated enormous discontent and kept the nation's economic mobilization policies and programs in a state of constant turmoil.

"Military necessity" became the chief rationale for wartime policies. This resulted in campaigns that were at best peculiar, at worst irresponsible. The struggle over national service legislation and reconversion policies will illustrate the point. Once the nation entered the war, a group of elite, World War I Plattsburg enthusiasts launched a drive on behalf of an extremely class-oriented national service legislation, designed to draft the nation's civilian working population. Such bills were before Congress throughout the war years. But no labor shortages existed before mid-1943 and thereafter they were only local in nature. Consequently, to justify these schemes, blatant or subtle arguments had to be made to the effect that labor was lazy, disruptive, unconcerned, unpatriotic, or uninformed, despite the fact that practically every family had close relatives in the armed services. Business never supported the national service movement. Instead, it remained silent and allowed its civilian and military proponents to carry the burden, which served to keep organized labor and spokespersons for the masses on the defensive.

A similar situation occurred with regard to the battle over reconversion. By 1944 the War Production Board had begun preparations for the transition to peace, but the big corporations and the military blocked the way: the former because they wanted to protect their postwar market positions by keeping all or most manufacturers restricted to wartime production or controls until hostilities ceased; and the latter because it feared that a stampede for civilian production would deprive it of munitions and supplies. As with national service, squelching efforts on behalf of reconversion required playing fast and loose with the truth and portraying the populace as untrustworthy. The genuine need was to calm the public's legitimate fear of a postwar depression, which had been predicted by many. Yet, little or no planning for reconversion took place, a situation that led to chaotic economic and political conditions once the war was over.

Growing public disillusionment over the way the home front was being governed accounts in part for the declining vote for Roosevelt and the Democratic Party. Additionally, polls point up the fact that public opinion was not only confused and contradictory during the war, but also that it had a callous, selfish, and uncaring streak. These developments were hardly surprising. When the president told the nation in December 1943 that "Dr. New Deal" had to give way to "Dr. Win-the-War," he was declaring that either the war was more important than reform or was incompatible with it. Either interpretation was disturbing and bound to cause worry and to generate resentment among broad segments of the public.

The president not only neglected or downgraded reform, he also failed to establish adequate controls over the military as the agents of war. That was the case despite the fact that Roosevelt had the experience, the opportunity, and the interest in doing so that Wilson, for example, never had. From 1939 to 1942, the president appeared to be going in the direction of establishing a war council that could have presented him and the nation with a balanced executive advisory or policy group for the entire war effort. That never materialized, but Roosevelt did create the Joint Chiefs of Staff in 1942, which gave the armed services direct access to the president at the expense of the departmental secretaries. This situation at best reflected the realities of power. The secretaries of war and navy during World War II played a less significant role than had their World War I predecessors. Roosevelt, curiously, had allowed the civilian leadership of both the War and Navy Departments to wither or disintegrate during the critical years in the late 1930s. In 1940, he finally moved to make the political appointments of Stimson and Frank Knox. The latter was a nonentity whom the entire navy and Navy Department structure simply circumvented and who forced the president repeatedly to intervene in departmental affairs to keep the navy brass from taking over civilian roles.[10] Stimson had much more experience in military affairs than Knox and he knew what was going on, but he was not a strong secretary of war. He viewed his role as that of protecting the army from civilians and running interference for it, instead of acting as a cabinet member directing his department as part of a larger administrative whole.

At the conclusion of World War II, the military had achieved a role, power, and prestige unequaled in American life. In contrast to the past, all of the services had confidently planned for their futures. The plans of the army and navy anticipated Cold War roles for themselves, but they were not grandiose. The Army Air Force, however, displayed a daring ruthlessness. It proposed a dominant role for itself in the postwar military system and then tailored a foreign policy and strategy to back up its conclusions. In the process, the air force selectively used the findings of the U.S. Strategic Bombing Survey that it had sponsored, and it insisted that supporting a massive aircraft industry was one more reason or even reason enough for maintaining a huge air fleet.[11] This aggressiveness stemmed in part from the fact that the air force was a new military arm that had been born of technology and was unhampered by the traditions and restraints of the past. More than the other branches, it was heavily dependent upon civilians and had readily turned to them before and during the Second World War. These civilians at times turned out to be more prone to taking extreme actions than the military itself, as the terror- and fire-bombing of civilian targets and the ultimate use of nuclear weapons established.[12] This new military arm had a substantial lead in advancing its cause over the more traditional fighting forces and compelled the latter to emulate it in imagination and aggressiveness in order to thrive, let alone survive, in the postwar world.

The Roosevelt administration had squandered the unequaled chance just before and during World War II for creating a fully answerable military system. By comparison, the Truman administration lacked the conditions, the prestige, and the experience of its predecessor for achieving that goal either when military budgets declined from 1946 to 1950 or when they began to rise dramatically thereafter.

Some business leaders and their allies in organizations such as the Council of Economic Development perceived that the armed services had achieved dangerous levels of power in America and that continued spending for the military at high levels could threaten the nation's economic and political systems. More traditional business groups, like the Chamber of Commerce, were not so concerned, although outside of the aircraft, shipbuilding, and machine tools industries, industry in general favored quick demobilization and did not look to huge military budgets to ensure prosperity. Whatever the views of the industrial-financial community, it failed to act in a way that could lead to more effective controls over the armed services. This helped to create a vacuum of leadership in an area with potentially grave consequences for the future.

The crusading zeal of World War I was not experienced during the Second World War itself. However, in the postwar years, as Cold War tensions mounted, an ideological crusade was unloosed that exceeded the earlier one in length and abusiveness. Passions may not have been stronger, but much else had changed. The nation's global reach and its institutions for war had ushered in a new stage of militarization.

From 1914 through 1945, war and preparation for it had a much greater impact upon society than they had had during any comparable period up to that time in American history. At the beginning of the period, the nation was unprepared to fight a modern war in political, economic, or military terms. Its efforts during the First World War were stumbling and uncertain. Because of the interwar planning, the United States was able to mobilize its might for the Second World War in an impressive although not flawless manner. At the end of hostilities, the power and influence of the military in American society had reached unequalled levels, which would be difficult to diminish.

The effects of militarization inevitably extended beyond warfare capabilities. Economic mobilization for the two world wars enhanced elitist power in the society. This acted to strengthen the opposition of the Neo-Jeffersonian reformers to war and preparedness and produced a series of investigations about the conduct and consequences of warfare that were remarkably revealing and astute. The Nye Committee stands out in this regard. Furthermore, warfare weakened democracy by increasing intolerance. War and revolution have fed one another abroad in the twentieth century and that reality came home during and after both world wars in the form of red scares. An elite devoted to a world of order tended to further rather than diminish these passions among a population exceptionally susceptible to ideological crusades.

The era of the Great Wars paved the way for but did not create the national security state of today. After the Second World War, the nation's fighting forces remained larger than ever before in American "peacetime" history. And the Roosevelt administration had passed up the last and best opportunity in the interwar years and at the outset of the Second World War for creating a fully accountable military system. As long as armed services' budgets continued to decline, then the nation could still rely on restricting military influence upon the society by keeping its size within reasonable range. Once military spending began to escalate rapidly from 1950 onward, the nation simply lacked the policies, the institutional structures, the traditions, and the experience for controlling its war machine. The voice of the armed services would grow in the formulation of foreign policy, the military's influence would become pervasive throughout society, and various industries, whole communities, and entire regions would become economically dependent upon military spending for their prosperity, even their existence. Once that occurred, America would become a warfare state. The years between 1914 and 1945 did not make that result inevitable. However, all of the developments, trends, and planning of those years almost ensured such an outcome, if massive spending for war continued over any length of time. That, after all, was the primary message of the Special Committee Investigating the Munitions Industry—the Nye Committee.

Chapter 4
The Militarization of Europe, 1914–1945

MICHAEL GEYER

THE PRESENT MILITARIZATION of Europe arose out of the collapse of nineteenth-century authoritarian and paternalistic militarism in World War I.[1] During the interwar years the pervasive economic, cultural, and psychological mobilization for war became a central feature for organizing European societies and their interactions. It was also during these years that Europe saw the rise of integral nationalism—mass-based and ideologically-driven nationalism—which was a prelude to the societal organization of violence and the militant ideologies that have been characteristic of the supranationalism of the Cold War. Ever since, moreover, we cannot think of Europe without considering the omnipresent and indiscriminate threat of extermination. Even the collective amnesia of affluence does not hide the scars of the violent origins and the militarized present of contemporary European society.[2]

This European condition runs counter to the expectations of virtually all social theorists of the nineteenth century, who either predicted an age of industry and peace or one of revolution and subsequent justice and order.[3] However, from the vantage point of the late twentieth century, the era between 1815 and 1914 appears to have been more an exception than the norm. It was an age in which a European peace was "invented" after the cataclysms of the Napoleonic era and the preceding age of state-led violence. This peace was subsequently lost in an age of mass mobilization for war. None of the social classes and none of the social ideas to whose rise the spread of peace had been attributed in the nineteenth century had achieved its presumed *raison d'être*. In the twentieth century the forces of production and destruction have grown simultaneously, and if the former has eradicated abject poverty in Europe, the latter has eliminated peace. Liberalism, nationalism, and socialism have all had their day, but all of these once potent ideologies of peace have also contained the seeds of war. Have we then begun to speak of nonviolence, because we have lost a notion of society and international order at peace?

There have been throughout the twentieth century social movements that have raised the specter of peace amidst military preparations, even if they have been unable to stem the tide. There is hope, even if there is no coherent practice, that the current crisis of European and world order may indeed initiate a process of demilitarization—an active reversal of the economic, cultural, and political mobilization for war. But there is also room for doubt. In the United States, pessimism and disarray reign supreme among advocates of peace. Current European antimilitarism tends to seek its opponents everywhere except in the present state of Europe and its origins—the self-initiated militarization of Europe in and after World War I. Unconscious of its own past, contemporary antimilitarism is unable to understand the present. It avoids self-analysis by turning its attention to the penultimate history of German militarism in the eighteenth and nineteenth centuries or to the impact of the "bomb culture" of the United States and the warfare state of the Soviet Union and their impact on the world. But the former preceded the militarization of Europe and the latter resulted from it. The European process of militarization, as it originated in World War I and was worked out during the interwar years, is the hinge that links them.

The boundaries of this period of European militarization are clearly circumscribed. German militarism triggered the outbreak of World War I and set in motion the process of European militarization. The inability to sustain this process or to demilitarize drew in the "world." World War II mobilized the United States and the Soviet Union with their own distinct and separate cultures of violence, which bore down on Europe.[4] Europe has remained unable to organize and reassert itself ever since. In the long history of the militarization of the world, the interwar years are the fulcrum and the turning point. The preparation and the threat of war not only became a central theme, but both spread to cover the world.

If we have come to live so comfortably and for so long with the current state of militarization, this is not the least the fault of the historians and social scientists who have averted their gaze from the militarization of Europe by locating militarism thoroughly in the distant past. How it was ever possible to identify twentieth-century militarization with the persistence of premodern (German) elites, while the Cold War and the threat of annihilation engulf the northern hemisphere and wars of liberation the southern one, must be the topic of another inquiry. It is, in any case, an intellectual outrage and a sign of the duplicity of an intellectual culture that has made an unchangeable past the enemy of a changeable present. What we can do here is briefly to characterize the nature of the assumptions that have so conveniently veiled our current problems.

Nineteenth-century militarism, which found its expression in the Prussian-German military state, is commonly understood as the predominance of martial values—encapsulated in the behavior, code of honor, and symbolism of a largely aristocratic officer corps—and military politics, reflected in the military state within the nation. Several intellectual traditions are fused in this image. The oldest one was concerned with the (im-)moral essence and identity of the "perpetual soldier," a longstanding metaphor. In the nineteenth century, this metaphor was historicized and linked to the rise of a military aristocracy or, more specifically, to the fate of the Junkers as they were transformed into a military caste and evolved into a professional body under the dual pressures of Prussian autocracy and the wars against Napoleon.[5] Schumpeter had this construct in mind when he juxtaposed, in the manner of Spencer, this caste, whose main function and skill was war, to the productive energies of the entrepreneur and of the rising industrial society.[6]

This image of a powerful military caste that was set against civil society was reinforced and popularized in the Enlightenment critique of military rituals of death and the other dangerous desires they held. This cultural critique was a critique of emotions, of the seductions of death and sex. Soldiers, much like women, were imagined to express dangerous desires that ran counter to the paternalistic morals of the bourgeoisie. Both threatened *Kultur,* industry, and the home, one from the outside and the other from within, and both undermined industry by the lure of primordial urges.[7] Both soldiers and women were supposedly susceptible to each other, and their body-oriented existences were prone to undermine the discipline of the spirit or, for that matter, of enterprise.

These cultural and social images had already receded in the interwar years. However, the charge of *Gesinnungsmilitarismus* (militaristic consciousness)[8] remained; that is, the military was presumed to have an extraordinary influence far beyond its actual social and political reach and was considered to be capable of affecting society at large. Societies, whose genuine interests and outlooks were presumed to be peaceful or "liberal," could succumb to militarism by taking on the *Gesinnung* (spirit), language, and behavior of military castes.[9] Civil society thus reflected the values and norms of a retrogressive class and gave an otherwise obsolete social group a new lease on life. This notion incorporated all the older connotations of seduction and danger. However, theorists continued to explain civil society's illicit affair with the military spirit in terms of political power. It was thought that the military mainly affected those who could not because of the weakness of their sex, or had not because of their emasculation, shown their political and social might.

In the United States similar notions have recently been put forward as a sanitized version of "secondary integration." Not accidentially, anxieties about the bourgeois control over death and sex were exchanged for the generalized fear and social mobilization of military ideologies. For this much had become obvious in the 1940s and 1950s: the dangers of militarism did not rest with an obsolete class; they lay in controlling the pervasive and encompassing social mobilization necessary to fight twentieth-century wars. The solution lay in containing the powers of the military as the organizers of violence, while harnessing their special knowledge.[10] The answer in postwar Europe consisted of turning officers into experts or professionals and in setting them against a "politicized" officer corps. Redefined as experts, the military could now be perceived as having distanced itself from the obsolete aristocracy. The transmogrification of Weber's critique of himself and the German bourgeoisie—minus Weber's fears about institutional domination—entered into the mainstream of historiography in the American discussion of the origins of German illiberalism[11] and justified at the same time "the professional soldier."[12]

From here this construct reentered a guilt-ridden German historiography and became the new orthodoxy of European history. Only one significant element was added, which gave the whole construct a seemingly "left" or Marxist glare. It was now argued that the stronger the challenge of industrial mass bourgeois society and the deeper the contradictions between the bourgeoisie and the proletariat, the closer the alliance became between the "old" and "new" elites and the more desperate the response and the more rabid and radical the search for expedients. The Third Reich could thus be explained as a result of the inability of German society to adjust to the modern, industrial age, and it fit this concept well if historians stressed the role of military desperados, another group of outcasts in a modern age, in the growing National Socialist movement.[13] Militarism was systematically associated with everyone and everything except the core constituencies of industrial society. Only outsiders and outcasts were "militarists."

This dramatization of German history served everyone. It allowed the western nations to consider themselves not just as victors, but as the flagbearers of modernity. After all, they had had their revolutions—or in America's case never had an aristocracy—and, hence, they were free of the vices of militarism, arming sensibly and professionally for the defense of democracy. Germans could get rid of their tainted past and Germany could join the march of progress toward a liberal, affluent society. After the critics of rearmament were silenced by the success of the German economy and a massive dose of Cold War ideology, one did not even need to wonder about the fact that never before in recent German history had there been so many soldiers stationed on German territory, never before had there been a similar danger of annihilation than existed in the Federal Republic and in the German Democratic Republic, and never before had militant ideologies shaped German society to the same degree. After all, the Junkers were driven from power and the problem of militarism had thus been categorically resolved.[14] The postwar armies were professional forces—therefore "modern" and nonmilitaristic.

This notion of German militarism is a comprehensive and powerful construct, but it does not serve a better understanding of the process of militarization in Europe between 1914 and 1945. Even if the assumptions made about the atavistic nature of militarism did fit the nineteenth century and even if they were appropriate for explaining the destabilization of Europe before 1914, they do not help us very much in decoding World War I and the interwar experience.

First we have to come to terms with the dissolution of the boundaries between civil society and the military organization, which at the turn of the century was still considered,[15] rightly or wrongly, to be one of the intrinsic and central features of European society, one of the cornerstones of the European state, and the origin of modern politics. It is not just that military men now transgressed these boundaries for the professional purpose of preparing and waging war, but that civilian leaders like Lloyd George and Clemenceau or for that matter Speer and Hitler, did the same, claiming control over the war effort from the military in order to better organize society for war. Civilian government and the military were no longer institutionally separated and both acted instrumentally in the pursuit of war. The preparation and the conduct of war became a comprehensive, managerial effort at national organization.[16]

Second, we have to make sense of the emergence of social and political groups, fascists as well as militant leftists, who did not just emulate military ways—who in fact more often than not rejected them—but based their identity on their military service for the nation, pursued the mobilization of society for war as social therapy, or, ultimately, aimed at the violent creation of new societies—societies whose *raison d'être* and identity were defined by war or the threat of it.[17] If commerce and production had been the hallmark of European civil society and the ownership of property its legal and ideological base, we now find increasingly the formation of social and national identities that were centered around (the threat of) violence and that presumed "essential" inequalities in the human race or irreducible differences between *Freund* and *Feind.* Social and national identities were defined in terms of hostile relationships.

Third, we have to account for the fact that Europe was the center of the process of militarization, but it was neither able to sustain that process on its own nor to cease hostilities before it had dragged in the world. Europe was the source of the militarization of the world, but also one of its victims. If one might argue that the capability of drawing in the world assured victory in World War I, one must add that in World War II victory had to be mortgaged in order to fight war. European militarization thus set in motion not only a militarization of the world, but made Europe its first and main casualty. Though Europe had gained its predominance in the world through its expansion of production and destruction, it lost that predominance due to its inability to control the global tensions that the process of militarization engendered.[18]

It is here that we can begin to see both the intrinsic importance of the process of militarization that took place between 1914 and 1945 for European affairs and its centrality for the transformation of Europe and the world. This was no longer old-style militarism. "Professional soldiers" and "managerial" government set in motion the militarization of Europe and militarized social movements gave their efforts legitimacy. The result was the diminution of Europe and the rise of global processes of militarization, in which Europe remained the single most highly militarized region of the world but no longer controlled its own destiny.

The crucial transition from the nineteenth to the twentieth century consisted in the end of the European capability to limit intra- and interstate violence and to contain it in the (European) state system.

As a result, the old distinctions between the military and civil society, war and peace, destruction and production, and the notion of militarism that built on them fell apart. We must draw new distinctions—distinctions that rely less on the formal characteristics of military institutions and military castes and more on the nature of professionally organized domination and subordination within Europe and of Europe in relation to the world.

Throughout the nineteenth century, military elites, confident of their craft, had learnt to put a premium on limiting and thus controlling war.[19] They expected that wars would end in a brief sharp clash, and the civilians seconded them in this notion. Notwithstanding militaristic sentiments, both elements considered war as an undue but occasionally necessary interruption of civilian life and pursued war and its preparation in this sense. In 1914 they looked forward to war as a highly risky but ultimately beneficial undertaking. War would resolve the costly stalemate of European international society and allow, after appropriate adjustments to be worked out in peace negotiations, a return to the predominant business of the day: shoring up empires, defending autocracy as well as parliamentary governments against their many detractors, and most of all, furthering the production and commerce that had made Europe the center of the world.[20] Others hoped that this war would cleanse the air, settle social tensions, and prove once again the manliness of men.[21] Yet others—and they were the ones who actually triggered off the calamity—anticipated that behind the shield of the European stalemate, the obstreperous people of the Balkans would be taught a lesson and their struggle for national liberation could be subdued.[22] This war, in any case, was not started by outsiders. It was set in motion by mainstream politicians.[23]

However, it was started in the firm expectation that the recourse to weapons could resolve the problems that had accumulated during the peace and would allow a return to peace—a peace that rested on the implicit consensus among European elites to contain rivalries between them in favor of domestic stability and their control over the global expansion of commerce. The fact that mainstream politics was wrong, that German governmental and military elites escalated, for hotly debated reasons, a Balkan incident into an all-out European war, and that this war could not be brought to a quick conclusion should not detract from the unanimity of outlook, which reflected a century-old experience with containing intra-European violence for the benefit of assuring global domination. Despite the arms race after the turn of the century and despite the growth of a new, "integral" nationalism, all European governments and their military institutions were geared toward containing war. The difference was that some considered war more feasible than others. None, however, foresaw the war that came about.

None of the armies developed the crushing dynamics that should have overwhelmed the enemies and forced them to sue for peace. This war was long and it laid bare the social arrangements on which the nineteenth-century European order had rested. It heightened tensions between social classes faster than it could recreate new social alliances; it destroyed manliness more radically than it was able to recreate images of men; and rather than teaching subordinate people a lesson, it strengthened their drive toward independence. In other words, the key arrangements that had sustained the European states and civil societies, allowed for the containment of violence, and underwritten European predominance in the world were challenged when the crucial prerequisite for these arrangements, the ability to limit the "lateral" competition between European nations, gave way.[24] What happened has been debated ever since, and it is best to uncover the layers of this development one by one.

This was first and foremost a new and unexpected kind of war, what General Ludendorff called a "totalitarian war" in what Eli Halevy designated as the "era of tyrannies." Several unique features of this war are commonly recognized. First, World War I was fought with an ever higher intensity of destruction, turning soldiers into moles and heroes into skilled workers of mechanized destruction. Combat zones began to extend backward in the first air bombardments, which together with the allied naval blockade and German submarine warfare dissolved the differences between soldiers and civilians, turning whole nations into armed camps. Second, while the actual combat zones remained narrow, the front lines spanned areas unimaginably far into Galicia and Russia and even into Palestine and Mesopotamia, where Indian and Ottoman troops fought against the ravages of disease as much as against each other—the first modern Third World war. Third, warfare now mobilized the whole nation, breaking down institutional barriers between the military and civil society. The armies at the front became mere conduits through which the nations poured their soldiers, their resources, as well as their hatred and prejudices. Fourth, economic and military mobilization was complemented by pervasive mind control, censorship, the criminalization of opponents of the war, the internment of enemy civilians, as well as state-sponsored patriotic mobilization. The intensification of destruction, the expansion of warfare into an intercontinental dimension, and the organization of all national resources for war were the main military characteristics of this war.[25] Together these elements constituted a new mode of organizing destruction. While it is easy enough to describe its elements on a phenomenal level, it is altogether more difficult to figure out its peculiar logic, which ran against the grain of older notions of militarism. For the source of this new mode of destruction was not "the other," but civil society itself. How can we understand war and its preparation under these conditions?

It is not surprising that under the impact of all-encompassing national efforts, we form the image of a cataclysmic war—of war either as natural disaster or, more in tune with the spirit of the age, as a huge, anonymous machine that sucked up ever more materiel and ever more men only to spit them out as weapons and soldiers going into almost certain destruction; we see war as an (industrial) dance of death engineered by a giant mechanism that subordinated whole nations under the imperatives of war mobilization.[26] And this was a war of the masses who did not cease to fight for four long years. It brought to the fore the great traditions of mass organization of the nineteenth century, leaving behind political and economic liberalism as well as the idealism of *Bildung* and *Kultur*—individualistic ideologies of a rapidly vanishing past. War also expanded the "iron cage" of institutional domination that Weber had dreaded so much, because it advanced more total and more comprehensive forms of bureaucratized mass domination. The latter organized society along functional lines and individuals according to their use value in destruction.

And yet, as much as we have to acknowledge the powerful thrust of state-centered national mobilization that reached the remotest corners of the nations and the world, the image of war as an overwhelming apparatus of destruction, as an industrialized Leviathan, is insufficient—in fact, it is counterproductive for understanding the process of militarization. It is true that national mobilization broke down entrenched divisions between the military and civil society, but this was not an ever-expanding process of technocratic domination that overwhelmed everyone and everything. This view is, at best, a mere negative dialectic, which points to a collapse of civil society but neither explains nor comprehends what came after its fall. It is a mystification of the fact that militarization originated in civil society, instead of being imposed on it.

Let us approach this problem step by step by first identifying the cross-currents of militarization. What is generally called the "transformation of war" in World War I consisted of competing and often outright antagonistic processes. On the one hand, we can observe an increasing active or passive participation of society in war through (1) the nationalization of military mobilization across the old civil-military divide and (2) the incipient "socialization" of danger that no longer separated soldiers from civilians. It is one thing to stress the dissolution of civil-military boundaries. It is another to see that this national mobilization of society could and did become a source of social self-assertion.[27] It raised questions about the status of mass participation in the organization of the war effort and national politics as well as about social identity and cohesion, and it challenged existing forms of political subordination and social deference. It broke down old modes of domination and created new ones; it recast class, gender, and ethnic identities. On the other hand, we find the growth of organizational and managerial power over society and its resources in (3) the industrialization and mass production of the means of destruction that passed the control over weapons to industry and (4) the managerial reorganization of the preparation for war and the military use of force. This reorganization process increased the capabilities of state control over society and set in motion contests over the "site" of that control—in the military, the "government," or in industry.[28] Another aspect of the same process is more difficult to grasp but perhaps even more important. The intensification of institutional domination pried open the old linkages between military and industrial institutions and their social sources of legitimacy (the bourgeoisie and industry, or military and aristocratic elites) and it opened cleavages between the institutional powers of industry and the military and the formation of social identities. Old entrepreneurial values and the new corporate ethos drifted apart, much as did the traditional fighting spirit and the new military organization.[29]

In short, World War I forced into the open two basic and contradictory processes that shaped the twentieth century: the problem of mass participation in national affairs on the one hand and the institutional-professional organization of domination on the other. At the same time, it drove a wedge between processes of institution-building and social formation and dissolved the identity between the state and social elites. The rearrangement of the triangular relations between mass participation, institutional domination, and elite formation is the "site" of twentieth century processes of militarization. These were always embattled processes: they involve the densely negotiated rearrangement of power relations throughout society and the state.

The arrival of mass participation changed the nature of the political arena quite radically, completing and at the same time transcending a third development that had been inherited from the nineteenth century. In World War I, we see the final stages of the long process of the nationalization of politics within the European nation states. The main political struggles in World War I, in all nations, consisted of defining the reach and the limits of politics and the "sites" for mediation between mass participation and bureaucratic domination. As the long process of the nationalization of politics was completed, the central role of national politics in shaping the destiny of the nation was firmly established, even if national politics—at its pinnacle—was embattled and even if it immediately fragmented into a myriad of partial mediations in state, industry, and society and gave rise to "parastatal" agencies (organizations with powers like the state, but that are not a formal part of the state).[30] While this process has received considerable scholarly attention, it is less evident that we also begin to see at the same time an increasing dependence of this nationalized politics on a transnational and global organization of power in terms of the international organization of finance, commodities, and resources. This process clearly went beyond the mere reorganization of tacit arrangements between Europe and the world. The fact that nations were now forced to depend on resources beyond their boundaries posed challenges to European control over global integration and showed first and foremost the limits of the nationalization of politics in Europe.[31] None of the European nations was able to reproduce itself autonomously, let alone engage in internecine rivalries. The process of nationalization in Europe hinged on the global organization of power and the control over resources that that entailed. War raised the stakes of control and yet weakened the ability to exercise it.[32]

If the "transformation of war" was indeed a highly contradictory social process, who and what set it in motion and who and what kept it going? Again, this question is easily answered, as long as we assume—in the tradition of the nineteenth century—that the preparation and the use of force are organized along a clearly delineated divide between the military and civil society. For under these conditions we can argue, albeit in somewhat abbreviated form, that the military (and those social forces that sided with or controlled the military) set the whole process in motion and kept it going.[33] However, this view does not answer the question of what happens if the divide between the military and civil society no longer exists and if the preparation and the use of force becomes a comprehensive and encompassing national and even international undertaking; and if society is not just a victim of princely or elite conspiracies but an active participant in the preparation for and the pursuit of war.

This most immediate response is not entirely wrong, though. There is a good reason why the preparation and the use of force are so commonly attributed to some conspiratorial elites or is seen as a natural force or *fatum* that hits societies, subordinates them under its imperatives, and equalizes them in the process of subordination. This turn to "hyperreal" actors reflects a genuine experience of war in the twentieth century.[34] War is everywhere and it engulfs everybody. War comes from somewhere "out there"—literally and figuratively—and has an "impact" on society.[35] War seems to envelop society and all its partial activities and to transcend it. War thus becomes a metahistorical force. Everyone participates, but no one takes responsibility. War—its purpose and pursuit—is "out of control."[36] This mystification is a direct response to the fact that war has lost its nineteenth-century quality of being the "other," subordinated and fenced in by civil society. As an interpretative device, however, it is a sign of our helplessness in the face of the fact that civil society itself originates and perpetuates war.

In order to avoid these mystifications we must move from the delineation of the "sites" of militarization to the analytic recovery of the "process" of militarization. First of all, we have to recall that preparing for and waging war is always a social process in which resources and "man"-power are appropriated, transformed into means of destruction (soldiers and weapons), and spent in the effort to subordinate an enemy. As long as the process is segregated from the activities of civil society, old definitions of "militarism" apply. War in the twentieth century, however, loses its quality of coming from the "other." War now becomes part and parcel of the formation of "civil" society—a process that was foreshadowed at the periphery as, for example, in the Civil War in the United States.[37] Nineteenth-century militarism, the uneasy relations between a military state/caste and civil society/government, both grounded in very different principles and limiting each other, gives way to an encompassing process of violent social formation that we call militarization. Militarization is the contradictory and tense social process in which civil society organizes itself for the production of violence.[38] Hence, the process of militarization does not "interrupt" or "disrupt," but it compresses processes of community and nation-building under the dictates of scarcity; for war entails the destruction of values, goods, services, and human beings.

Hence, militarization is often described as a process of impoverishment, immiseration, and of loss for everyone. To be sure, the losses in twentieth-century European wars were extraordinary, but so were the reproductive capabilities of societies under extreme duress and so was their subsequent ability to forget. War as Armageddon does not sufficiently describe the balancing act at the brink of the abyss. Nor is the notion helpful that war was or is a productive force in the last instance.[39] It can be productive and destructive at the same time. What matters is that losses and benefits are distributed unequally. The struggle over who bears these costs, nationally and internationally, is the substance of the politics of militarization. These contestations, however, are not merely matters of cost-benefit calculations—a struggle over the distribution of "resources," as if institutional rationality shaped the process. Rather, it is a social conflict in which the identity and cohesion of social groups is rearranged. Why should men "go to war" and not women? What happens when a whole generation of men is killed? What effect does the retooling of industry have on class formation? What place do "colored" troops and coolies have in war and, more generally, what is the role of the colonial "resource" base? In short, the politics of militarization consists in the rearrangement of relations of class, gender, and ethnicity, and the ideologies that guide them. What is true for the organization of production as a universalizing process also applies to the organization of destruction, once it was set free from its nineteenth-century constraints.

It is, indeed, not war or "militarization" that organizes society, but society that organizes itself in and for war. In mobilizing people and resources for war, societies remake themselves and their social-political orders for the purpose of organizing destruction. This at a very profound level is the meaning of "militarization" in twentieth-century Europe. This is also the appropriate entry point into the problem of the origins of militarization. If militarization is a process of social reconstruction, its origins cannot be found in the competition over interests, but in the struggle over social and national indentities. With the collapse of the boundaries between civil society and the military, there was no longer a place for *Gesinnungsmilitarismus,* the transfer of military values into civil society; for civil society has reconstituted itself on the basis of violence, that is, in the pursuit of war. In this it differs profoundly from nineteenth-century militarism, and in this lies the difficulty of defining peace in an age of violence. The organization of destruction has become a "primary" means of social integration.

Once we have established the nature of the process of militarization, we may proceed to distinguish between "militarization" and a new "militarism" that fits the age of total war and should not be confounded with nineteenth-century militarism. The process of militarization has its own gradations to the extreme. If militarization reorders social arrangements and defines social identities, the new militarism aims at the creation of a social order that lives off violent subordination and the militant purification of the dominant "race." There is a small but decisive step from militarization to militarism under the aegis of total war. Militarization turns into militarism if and when the maintenance or reproduction of the social order is made dependent on war—if war is not just the continuation of social organization by other means, but if war becomes the very basis of social organization; that is, if societies live off war or its preparation and propagation either economically, politically, or culturally.[40] In view of the nineteenth-century division between war and peace, between military and civil society, this distinction may look like hairsplitting. But we must remember: this is a violent age and in this age the difference between militarization and militarism matters.

The encompassing and comprehensive mobilization of the nation for war was a common feature of all the major belligerents in World War I. The main challenges were the same everywhere: mobilizing people for war while limiting their participation; expanding the resource base without forsaking national autonomy; and mobilizing for war without setting off a process of militarization. In order to achieve this end, all the nations resorted to a tangled web of compulsion and suasion, developed national forms of management, and undertook extraordinary efforts to shelter their prewar social and power relations. They all proceeded reluctantly onto the course of national mobilization. Their initial impulse was conservative—to fight a war *à outrance* in order to preserve prewar society. But in 1915–16—in the context of the various national munitions crises—all the nations involved considered peace more dangerous than the pursuit of war—an indication of the collapse of a European elite consensus and a testimony to the fear about domestic reprisals if war ended short of victory. However, the social arrangements that were necessary to accommodate this national mobilization drove all nations onto the path of militarization. It was a contested process, to be sure, and just as the social arrangements for national mobilization varied, the process of militarization pointed in different directions as well.

France was undoubtedly the most radical state in espousing all-out mobilization. However, even at the height of Clemenceau's rule, this process did not lead to an effective centralization of power in the state, but rather to a very contradictory situation in which large industries formed an increasingly coherent power block, social control and political power devolved to provincial *fonctionnaires,* and national autonomy was lost to an increasingly international organization of resources that were controlled from abroad. French national politics desperately tried to straddle all three developments. The failure to achieve a new consensus set up the social and political tensions that dominated the interwar years.

The technocratic schemes for organizing production that strengthened French managerial elites and central state agencies in the common pursuit of efficiency-oriented mass production and the centralized management of the labor pool has been elevated as the key feature of the French war effort.[41] Suffice it to say that corporate industries succeeded in establishing themselves as a powerful block, sometimes in competition, but mostly in alliance with state bureaucracies and even the military. As celebrated as these efforts at national mass production were in transforming the pattern of French industrialization and in ridding state and managerial cadres of what they considered the debilitating effects of popular politics, these centralizing schemes were sorely tested as wartime scarcities grew and infrastructural problems came to the fore. As important as the centralization and modernization of French industry was, it was only one element in a more complex rearrangement of power relations that shaped the decisions over organizing the nation for war.

The control of increasingly scarce resources and of capital and, hence, indirectly of the whole French war effort shifted to the international level, where the Allied Maritime Transport Council, foremost among interallied organizations, effectively began to decide during the last two years of the war what and how much France could produce and to what degree French civil and military demands were being satisfied.[42] As a result, the French nation never faced the full effects of the destruction of resources and commodities, even though it payed the highest human toll of all belligerents—as befitted a country now dependent on its allies. France had become by 1918 a platform for destruction that was supplied, organized, and maintained from abroad. This had the result that one potent source of militarization was externalized. National autonomy, however, was reduced.[43] While France went through a managerial revolution and moved away from the old entrepreneurship to develop a more corporate form of big business, it lost its independent position as a national producer. These conditions shaped one of the faultlines of conflict in the course of French militarization. The externally maintained modernization of industry forced the corporate sector into a balancing act between the international economy and national interests. The consolidation of corporate industries in France depended on war and war-induced conditions.

If French industrial and managerial elites were unable to control the commanding heights of an internationalized process of militarization, national control over the mobilization of society was maintained and expanded, though by means of yet another trade-off. One is, first of all, struck by the fact that due to the peculiarities of French colonialism, colonial resources and manpower were mobilized only very slowly.[44] The French countryside on the other hand became the main target for mobilizing manpower for the military and industry and it was bled white in the process. Whereas the number of colonial troops grew slowly and hesitantly and Chinese and Vietnamese labor was limited to digging trenches for allied soldiers, the French countryside provided the bulk of the soldiers and of the unskilled laborers who worked in the new and rapidly expanding war industries.[45] The female labor force took up the slack in the countryside, working less in factories than in home industries and especially in agriculture, though they could also be found in growing numbers in the *usines de guerre.*[46] A growing group of urban and provincial *fonctionnaires* organized the process of mobilization. This process of mobilizing the "backward linkages" catapulted the countryside into the center of national political conflict and established the second faultline in the militarization of France.

The military, industrial, and agricultural workforce remained remarkably mute throughout the war.[47] This was only partly due to repression or to the drafting of labor leaders. More important, it seems, was the exploitation of both the gender division of labor and the enduring split between city and countryside. These struggles found their political expression in the rhetoric of the *nation armée,* which was appropriated not by corporate managers, but by provincial elites. The recovery of this metaphor strengthened the status and the political power of the provincial bourgeoisie, which made the war into its own national cause and expanded its control over the *pays* and the social organization of power in the course of wartime mobilization.[48] Provincial elites and *fonctionnaires* thus managed a course of mobilization while limiting participation. The resurgence of the provincial bourgeoisie was the second pillar on which the French war effort was built. That class played a decisive role in controlling the process of social mobilization and in the concomitant formation of a class of patriotic soldiers and laborers who were most important prerequisities for Clemenceau's rule. The status and political power of the provincial bourgeosie was tied to its role in wartime mobilization and its preservation in the collective memory of the war. The countryside became the incarnation of the national spirit. French collective memory of the war was quickly usurped by provincial society, which identified itself with the *nation armée*. Provincial society juxtaposed this mythical nation both to socialist and technocratic forces and legitimated provincial social arrangements, based on the subordination of peasants and landed labor and on the preservation of patriarchy.[49]

The wartime service for the nation elevated the provincial bourgeoisie and successfully demobilized the countryside. The two aspects of this French process of militarization should not be torn apart. War became a means to defend the *pays* against the city: it prevented the linkage between rural and urban discontent and legitimated provincial social relations. At the same time, this was a "republican" militarization that turned against urban politics and corporate industry and, most of all, opposed industrial war. If this was "pacifism," it was a very strange brand of pacifism indeed. The provincial bourgeoisie turned against war, while defending and legitimating its social status by reference to its "national service" during the war. It shaped militarized social identities while opposing corporate hegemony. This is only surprising if we forget that the struggle over militarization was, indeed, a triangular contest rather than a collision between masses and elites.

The politics of militarization left a deep imprint on France. France became more provincial in the course of the war, while its technocratically managed industries became increasingly dependent on the forward organization of production, raw material supplies, transport, and finance out of the United States and Great Britain. France, in other words, externalized resource mobilization and at the same time militarized the countryside. Wartime mobilization put into place powerful, if competing, myths about French society that highlighted the dual poles of wartime and postwar hegemony in France: the notion of a republican France nourished by provincial *fonctionnaires* and the notion of a technocratic, "modern" France. Both together submerged the wartime experience of the rural population and of workers and they met in the search for autonomy for France, whose prerequisite was victory and hegemony over the continent. The social organization of the war effort established the parameters for postwar politics,[50] which remained wedded to the violent conflagration that had catapulted corporate industry and a militarized provincial bourgeoisie into the forefront of national politics in the first place.

The British victory was little less borrowed than the French one, though the kind of loans it depended on were different. Great Britain expended much more capital than France to finance the war, depleting its imperial assets in the process. But its per capita casualties were not only considerably lower than those of France, they were also matched by those of Canada and Australia. Great Britain resolutely drew upon the resources of its imperial domain in order to wage war.[51] As in France, the process of militarization was externalized, but British elites succeeded in maintaining control over it.

It proved to be exceedingly difficult to discard the restraints on war in a nation whose main business was the maintenance of empire and whose administrative and military state-building had occurred in the colonies rather than at home. However, once the predominance of imperial sentiments and of the social arrangements that maintained them were broken under the imminent threat of the collapse of the British war effort in Europe, Great Britain consolidated a national militarizing alliance that centered around three elements. A tight fiscal control over the war effort turned British mobilization of manpower and resources into the most state-centered and efficient mobilization in all of Europe. A dramatic expansion of production was achieved through the retooling of industry during the war and amidst growing scarcities, with dire consequences for an entrenched labor aristocracy. And a system of social services contained this expansion of production within clear-cut class, gender, and ethnic lines and was geared to maintaining prewar social identities.

The success of British wartime mobilization depended on the ability to unite two competing elite groups, national manufacturers and the financiers of The City, in order to overcome the munitions crisis of 1915. Acting together they managed to control the transformation of British domestic industries toward capital-intensive production. They relied on their ability to expand colonial extractive industries with the help of converted imperial assets and thus they moved the economy toward a new division of labor between domestic capital-intensive production, the dispersed production of basic goods, and dependent, colonial raw material extraction. This conversion contained at its core a transformation of the imperial structure of British society and industry.[52] It limited dependence on the United States but relied on the compliance of the dominions and colonies in support of Britain's war in Europe. Both the transformation of imperial ties and the retooling of industry required the rearrangement of imperial social relations.

The mobilization of dependent countries was contained at the cost of a growing independence in the dominions and of a strengthening of the state in the colonial countries. The wastage of Indian elite troops in Mesopotamia—which, incidentally, far outstripped the military blunders committed at Gallipoli—the rise of indigenous Indian production, and the simultaneous political mobilization in India, the key to the British imperial system, as well as the violent suppression of protest in the aftermath of the war were only the most prominent examples in a long chain of mobilizations and subsequent pacification campaigns waged during the interwar years.[53] The dominions' contribution of human and material resources, which equaled the contribution of Great Britain in relative importance, helped to diffuse the costs of the war further, but it also led to the renegotiation of imperial ties, as was most evident in the case of Canada.[54] The explosiveness of these conditions is exemplified by Ireland, where wartime militarization and ethnic-national mobilization could not be contained. Everywhere within the British sphere we can observe a devolution of power that followed in the wake of externalized militarization.

British labor was as deeply affected by the imperial transformation as were colonial and commonwealth elites. It is not by chance that British strikes far outnumbered those in any other European country except Russia.[55] These strikes are a testament to the fighting spirit of the old working class, but labor's inability to expand its class mobilization forward into the political realm of organizing war production and its inability to forge backward linkages to social unrest about living conditions, the tensions between a skilled (male) labor force and unskilled, "diluted," often female labor, and its inability to break out of the cocoon of working-class culture, blunted its effectiveness. If the old working class continued to hold control over the shop floors of the old industries, it lost out in the battle over a militarized process of restructuring British industry. It also lost out on the battlefront. The old working class seems to have carried the brunt of the British war effort, whereas the traditional recruits for the imperial British army, the rural and urban underclasses and the people of the Celtic fringe (including Ireland) fared relatively better, since these groups often did not fit the medical standards for conscription. The transformation of industry and conscription combined to rearrange social relations within the subordinate classes and undermined the hegemony of the imperial labor aristocracy.[56]

We can only fully gauge the far-reaching rearrangements of imperial social configurations, however, if we consider the disastrous consequences of wartime mobilization for the middle and upper classes in England. It is somewhat premature to make comparative judgements, but it seems that Great Britain was more reluctant than the continental belligerents were to broaden the recruitment base for officers and to substitute (lower-middle-class and even working-class) NCOs for junior officers, as was done in Germany.

Since officers—and particularly junior officers—suffered the relatively highest casualties of all ranks, the "well-to-do middle and upper classes . . . suffered disproportionally heavy losses."[57] And since overall Britain suffered fewer casualties than France and Germany, one must presume that officer casualties weighed disproportionally heavier than on the continent. Comparisons aside, the social base for the old imperial elites was radically depleted. If the German and French people as a whole faced a crisis of reproduction, this fate was limited to imperial elites in Great Britain.

And yet, the legacy of the British process of militarization was much more ambivalent than this seemingly clear-cut transformation of the imperial social and power structure indicates. It should be evident that the overwhelming urge of British politicians to return to "normalcy" was more than a thoughtless exercise in political futility. But it is not at all clear why the return to "normalcy" should have succeeded as it did, when the Lloyd George coalition collapsed shortly after the war and Great Britain as a result did not become a "home for heroes." This coalition, of course, had aimed at making wartime militarization into the basis of postwar politics and, in this sense, it acted in consonance with continental European politics,[58] whereas the effort of Conservatives and Labour politicians to recover the past set Great Britain apart. Great Britain in the 1920s was rather the exception than the norm, and the most likely explanation for this phenomenon was the absence of social alliances either for or against militarization. Despite far-reaching social rearrangements within Great Britain and the empire, British society did not form militarized identities.

It would be pleasant to think that the rejection of militarized identities was due to longstanding antimilitary sentiments and the British "liberal conscience." Two other elements, however, seemed to be more important. First, the process of externalized militarization and the very tightly class-bound internal process of militarization mediated against the formation of new identities. There was certainly no lack of proponents for and opponents to militarization. Colonial elites demanded autonomy, workers struck against rationalization, pacifists mobilized against war, middle- and upper-class women protested against the carnage, and social elites mourned a "lost generation" (of gentlemen), much as war veterans demanded entitlements for their service to the nation, new industries and dependent producers pleaded for the rearrangement of imperial policies, and writers and intellectuals considered war as social therapy. But the frictions between all these groups remained stronger than any sense of unity and the dispersion of the process of militarization rather increased than decreased fissions.[59] Segmentation preserved the hegemony of imperial elites, as hollow as this hegemony may have become if we consider the destruction of elite males and the rearrangement of imperial ties.

A second element is of equal importance, even if here we are on much thinner ice. Great Britain is not known for having developed a strong state, and yet it is the only country in which a tight fiscal control over the war effort was maintained and in which a remarkably efficient system of social services and of the distribution of scarce food supplies was established.[60] Both helped to contain the kind of civil unrest that proved to be so crucial in the destabilization of central Europe. (Its absence in Ireland or India may be noted in this context.) But the efficiency of this organized system of reproduction is not quite the whole story. Rather, the same system also underwrote the merger of middle- and upper-class voluntarism and state intervention, whose primary effect during the war was the reinforcement of class solidarities while limiting outright impoverishment. Efficiency was linked with the preservation of class identity.[61] This fusion of state and class politics in the process of militarization was costly because it worked both ways. Great Britain sent middle- and upper-class officers into death until the very end of the war and thus depleted the ranks of the middle and upper classes, but class identities and the system of class, ethnic, and gender relations, of domination and subordination, was preserved. Contrary to France or Germany, the class-based mobilization for or against the war did not spill over into the organization of the war effort—even if class relations were now increasingly mediated by the state. British wartime politics allowed for a high degree of mobilization without any concurrent politicization. State institutions retained a high degree of autonomy rather than becoming the battlefield for social mobilization.

The British Empire went through wrenching changes in the process of militarization, but basic class solidarities and identities remained remarkably intact. This distinguishes Great Britain from continental Europe (and, for that matter, from some of its colonial and commonwealth countries). Once the war had ended, British society left the war behind. Forgetting war or sentimentalizing it, however, neither obliterated the suffering nor could it undo the changes in the imperial structure. Both could be held off for some time, but they were to haunt Great Britain in the 1930s.

Contrary to France and Great Britain, Germany and for that matter Austria-Hungary had only one choice. They could either sue for peace or engage in an all-out effort at self-exploitation to win the war. Austria-Hungary tried both and failed in a maze of ethnic politics and bureaucratic infighting in which the German element now got the upper hand decisively. The incipient defeat of Germany dissolved the multiethnic empire, ironically despite the extraordinary military successes of the Austro-Hungarian armies in 1917–18.[62] Austria-Hungary, rather than Germany, should be juxtaposed to Great Britain, if we want to understand the possibilities and limits of segmented militarization. In Austria-Hungary this process failed, because elite consensus collapsed under the pressures of ethnic-national mobilization—first and foremost of German Austrians against everyone else—which successfully subverted the process of segmented militarization.

Imperial Germany chose war and all-out mobilization, aimed at the total organization of society and economy. As Ludendorff put it, the producers of weapons and the military strategists ruled this kind of warfare state as the supreme managers of the nation. However, intent and practice were two quite different things.[63] The technocratic dreams of a functional reorganization of society neither overcame the cleavages within German society nor evaded a relentless tug-of-war over control of the expansion of the means of coercion. The result was in many ways the opposite of what the military leadership and the industrialists had expected.

What was intended to be a destruction of politics in favor of nationalized and institutionalized domination in actual fact strengthened politics. From 1916 on the role of the civilian administration declined, but the role of parliament and especially of parastatal mediating agencies steadily increased.[64] If industry had hoped to gain control over labor, it ended up sharing power with trade unions in the allocation of labor and entering into negotiations over working conditions.[65] Power-sharing and other forms of collaboration could be extended into other sectors, but the conclusion remains the same. The deeper state and private institutions intervened in society, the more they were forced to make manifold arrangements over the process of militarization. While these negotiations clearly protected some groups more than others, they enhanced militarization overall. Hence, one must conclude that the militarization of German society proved to be a "success," due to an elaborate system of power-sharing and on-going negotiations over the process of militarization. Instead of an autocracy of a technocratic elite and of functional machines, we find the continuation and intensification of the institutionally mediated exploitation of German society for war.[66] If anything, it is not repression but this kind of "thick" mediation between highly organized social, industrial, and state institutions that distinguishes the German from the British and French processes of militarization. This only surprises us if we miss the crucial cue, which is a thorn in the flesh of German liberal historiography. The center of German politics, which was eventually to form the basis of the Weimar Republic, also formed the backbone of the German war effort. Despite intense conflicts over war aims, collaboration in order to share the control over the process of militarization was the key feature of the German process of militarization.

In entering into negotiations, parliamentary parties, trade unions, womens' organizations, and welfare officials supported the transformation of German society for purposes of expanding the organization of destruction, which could only prey on itself. There was no colonial hinterland and there was no United States to relieve the pressures of militarization. The more intense the efforts of destruction, the deeper went the cuts. Because war production and the mobilization of men and women increased dramatically in 1916–17 (even if they lagged behind official plans), and because there was no place and no race on whom to unload the costs, Germany faced a crisis of reproduction as it exploited itself in order to continue war—a crisis that took on catastrophic proportions, because collaboration worked so well. The problem was that there was less and less to mediate and the ever-smaller resource base was divided unequally.[67] While France and Great Britain faced crises of interallied power-sharing and struggles over subordination that pushed the process of militarization into the countryside or externalized it, Germany faced in dramatic ways a struggle for survival by a society that was ravaged by war.

It is in this crisis of reproduction that new social formations emerged that began to transcend prewar social identities: society began to organize itself along the issues of war and peace. On the one hand, we find the rise of a quickly growing militant movement, and on the other, we can observe the mobilization and organization of an antimilitarist, "peace" movement, both outflanking the organized nodules of the German war effort. While the latter figured prominently in the November revolution, the former is commonly underestimated. And yet both grew simultaneously and both reached their apex in 1918. Never before had antimilitarist sentiment been as strong as in 1918, but it was matched in every respect by prowar mobilization.[68] Both out-flanked efforts at power-sharing and collaboration, the thick mediation that involved the mobilization of people and resources in search of alternative strategies for Germany and in the struggle over social identity and hegemony. If the effort at social restructuring succumbed in November 1918, this was due to the incipient defeat of the German war machine.

The power of mobilization in the formation of new social identities and solidarities is instantly evident if we look at the social unrest in 1917–18 as a European-wide phenomenon. Only in those countries in which new social identities and new social alliances were based on wartime mobilization and the resistance against it (rather than on peacetime class alliances) did insubordination turn into revolution.[69] And only if we keep in mind the different practices of militarization do these revolutions make sense.

We may first notice the peculiar coincidence of "mutinies" in the French, Italian, and Russian armies.[70] These armies had one element in common. They all mobilized, for different reasons to be sure, rural and peasant armies and in pursuing war they dug deeper and deeper into the countryside. These rural armies, it seems, fought as long as the legitimacy of authority and as long as deference was maintained. Once both cracked open—due to the devastating blows of industrial war—they all resorted not so much to rebellion as to noncooperation and nonparticipation in the war.[71] The different paths of these insubordinations are startling as well. In France, authority was reconsolidated by George Clemenceau, whose politics reflected the republicanism of French provincial society, and by Pétain who appealed to the patriarchal values of the countryside in strengthening the authority of officers with openly antiurban and misogynist ideologies—shaping in this process the memory of the war as a war of republican defense of the patrimony.[72] In Italy, the legitimacy of the war was always more tenuous, but the Italian army mounted a violent campaign of repression in which a largely petit bourgeois officer corps subjugated southern peasant soldiers.[73] In both cases rebellion was subdued because the linkages between noncooperation at the front and the resistance against militarization in the countryside at home were cut. In Russia, on the other hand, this linkage was strengthened in the common mobilization for peace. Russian mobilization forged the separate struggles of soldiers, workers, peasants, nationalities, and women into a unified front—military insubordination turned into a revolution against war, against the pervasive and oppressive process of militarization.[74]

Neither is it by chance that Germany brought forth a very different revolution, while Britain had none. This had little to do with their respective structures of government. In Great Britain, resistance against militarization remained class-, gender-, and race-bound. Germany's revolution, in contrast, was fundamentally different from the Russian one because Germany fought a very different war. After having depleted the countryside in 1914–15, it conscripted workers and the petite bourgeoisie into the army and elevated NCOs to the level of junior officers. These military men did not launch a rebellion against authority, but they struck against war and its organizers. They made known in their demands that they fought the war and demanded a say in who commanded them and in how, when, and against whom war was fought.[75] This strike became general when it linked up with the parallel mobilization against war at home, which transcended class and gender divisions in a unified resistance against institutionalized domination, and in competition with the rising prowar movement.[76] The struggle of workers for direct participation and the military strike merged into revolution. In fact, they were actively combined in the resistance against the oppressive machinery of collaboration and collusion, and they were mediated in the food riots by women and in the insubordination of youth. These two groups effectively linked soldiers with workers in juxtaposing the preservation of family and social autonomy against institutionalized militarization.[77] To be sure, the German warfare state faced defeat in 1918 at the hands of the allies and this tipped the balance against the prowar mobilization, but in defeat the system of organized militarization was overthrown.

Nowhere, then, was militarization a matter of adapting to the "needs of war" or to new technologies of warfare.[78] Rather war was "made" in the encompassing processes of the social organization of destruction, much as in the nineteenth century "peace" had been made in the encompassing social organization of production. War is indeed nothing but a process of the construction of social relations and identities for the purpose of destruction. Most of all, war became a process of social organization and identity formation—rather than merely being its disruption.

Even if there were agreement on this point, as well as on the argument that this was a novel, distinctive, and European-wide phenomenon, can one truly say that Britain and France militarized? Did they not preserve their civil traditions much better than Germany? Did they not altogether escape the process of militarized mobilization? Was there not an essential difference between autocratic and democratic traditions?

There were significant differences to be sure, but I would put the essence of this difference somewhere else. After several decades of intense historiographic debate, it seems untenable to argue that autocracies were more repressive than democracies. It seems more true to say that there were differences in the actual practices of repression. And even if it is difficult to develop a measurement for the degree of militarization (which affected social groups unequally, in any case), it cannot be said that Germany militarized more radically or totally than did other nations, except when we fall prey to the lingering sentiment that somehow Germany was better organized and its people were more disciplined. It cannot even be said that the process of mediation over domination and subordination between social groups and between institutional powerholders and society was more highly developed in England or France than in Germany. Quite the opposite seems to be the case. There is one factor, though, which made a crucial difference. Both Great Britain and France blunted the domestic impact of militarization by militarizing their backward and forward linkages within their own countries and in relation to the "world." They mobilized along the class, ethnic, and gender divisions of labor within their empires and colonialized their countrysides in a process of militarization that mobilized men and women and excluded them from participation at the same time. These politics of exclusion and segmentation collapsed in Germany and for that matter in Austria-Hungary.

If we stressed the role of a crisis of reproduction in Germany, we must instantly add that it was not the geopolitical calamity it may appear. Rather, it was the consequence of choices that favored the continuation of war beyond the point the war could have been fought using the surplus accumulated during peacetime.

One could almost wish that this choice had been taken, as it is portrayed in much of German historiography—the premeditated and unscrupulous plunder of a small elite that was conditioned by centuries of militarism. But in fact the choice was made by managerial elites and military technocrats for instrumental reasons under rising popular pressures against and for war and the choice was condoned by the majority of political parties and all those functionaries who saw in it a chance to improve their position. It is this combination that sheds light on the rise of twentieth-century militarism. As befits the age of "masses," this militarism had a mass base from the very beginning. It is under the pressures of mass mobilization for war that militarism turned into a project of creating a society whose main *raison* was neither production nor civility, but war.

War did not end with the defeat of the central powers in 1918. Rather the social and national tensions of World War I spilled over into postwar domestic and international wars over the reorganization of Europe. These were new and unfamiliar kinds of wars for a Europe that had been used to and abused by wars for centuries—wars of militant social movements, imperial "pacification" campaigns in the context of the expansion of the imperial state, ethnic-national wars with genocidal tendencies, or class wars with tyrannical intent. While they petered out in the 1920s, a new and more permanent peaceful order for Europe was not to be found and it has eluded Europe ever since. The relations between nations and between social classes within nations remained in a state of suspended tension, which had only been contained temporarily by a semblance of international economic order. These tense stalemates did not survive the social, economic, and political turmoil of the 1930s and gave way to a new wave of mobilization and militarization of European nations. Their renewed preparations for war followed, once again, very different trajectories,[79] but they quickly engulfed Europe in another even more brutal war. The subsequent stabilization of a European order centered around military containment and deterrence.

It has been common to attribute the cause of this pernicious development to German elites or more generally to the unsettlement of German social conditions. This is true to a point. Germany was the aggressor in both wars, and the inability to reach a settlement over Germany has been at the core of the Cold War in Europe since 1945. But Germany was part and parcel of a European condition. It was the lynchpin of a European-wide development rather than its counterpoint.

Wars have been a permanent feature of European history, but at least in the nineteenth century wars had been contained and limited by a set of implicit understandings that shaped the European order.[80] These fell apart in World War I and this breaking unleashed a process of European militarization. The exact nature of this breaking needs further exploration, but two principles seem to be fairly obvious. The first implicit condition was the limitation of "lateral" tensions between European nations in a tenuous pan-European elite consensus in which the "Great Powers" ruled over small nations within Europe and extended their hold over the world. The second implicit understanding consisted in the internal limitation of war; the preparation for war and the use of force stopped short of reorganizing the nation for the purpose of war. In both cases, the preparation and the use of force depended on firmly entrenched peacetime structures of social organization and political order.[81] Both ceased to hem in war in the European crisis of the interwar years. At the end of the long process of the militarization of Europe, under the aegis of a Pax Americana and Pax Sovietica, the threat and the preparation for war held civil order in place. The fact that this latest militarization of Europe was controlled from its Atlantic and Eurasian fringes even if it was set in motion by internal European affairs only highlights the militarization that was already evident in World War I. Europe became the first victim of its own militarization.

While a careful analysis of European efforts to contain war and to erect new dykes against militarization after World War I must wait, this much can be said. Even in France and Great Britain, where elites retreated into empire as a counterpoint to domestic militarization,[82] World War I destroyed two of the most enduring metaphors about war and peace in Europe that had guided enlightened thinking ever since the eighteenth century: that nation-states could live in peace, once the nation-state had been firmly established as a principle of order in Europe; and that people are peaceful, while princes are belligerent. Both principles are linked by the identification of popular sovereignty with peace. They were at the core of the effort to establish a bourgeois order for Europe.

Much of European historiography during the past decades has centered around the effort to prove these principles and to disprove the existence of a process of militarization in Europe. This is the place where the construction of a German *Sonderweg* (the German difference) gained its significance. For as long as one could point to the "otherness" of Germany, one could uphold the validity of bourgeois principles for organizing Europe. This laudable intent unfortunately does not square with European development nor, for that matter, does it fit with the establishment of a Pax Americana and Pax Sovietica after World War II. The crux of the matter is not whether Germany was different, but lies rather in the very nature of this German difference. It was not princes but technocrats and professionals and it was not an old elite but mass movements for war that shaped the militarization of Germany after 1915–16—and these very same conditions are responsible for the continued militarization of Europe after 1945. They undermined the nineteenth-century bourgeois order for Europe. The German "difference," in other words, does not point back into the past, it points into the future. If England was the model for Europe in the nineteenth century, Germany played the same role for the twentieth. The defeat of Germany did not end, but only expanded and intensified the process of European militarization.

The key problem then is this: The effort to contain war by reestablishing a European elite consensus and by containing mass participation, the return to nineteenth-century normalcy, failed. All kinds of ingenious strategies were designed to limit war to an elite contest, to be fought by an underclass, by mercenaries, or by proxy forces, or to segregate and subordinate war to an international economic order. Seeckt, Liddell Hart, and De Gaulle were three of the more famous military champions of this course. Their efforts were complemented by diplomatic strategies to rebuild a European consensus around economic growth. These schemes were all rudely felled by unchecked competition and by mass participation in the propagation, preparation, and the use of force. In fact, this mass participation has emerged as the central aspect of militarization. Nowhere in Europe could war be fought against the "masses"—the French course in 1938–40 was exemplary in this respect. But the "masses" did fight. Whether we like it or not, key constituencies who were once supposed to assure peace mobilized for war—and, as a result, what we got were not calculated military confrontations for specific gains, but wars over "identity"; that is, wars in which whole societies defined themselves in opposition to a mortal "enemy."

If we study the abundant war literature since World War I,[83] we find a very curiously ambivalent attitude toward war. Even outspoken proponents of war abhorred its destruction and losses, and yet they embraced war as a social practice. For war was a central value in defining social status and the place of the nation in the world. Conversely, many opponents of war did not reject the "idea" of war, even if they opposed the barbarity of warfare. Thus, historians of Germany are puzzled by the phenomenon that the same people who favored rearmament[84] in the 1930s were stunned and shocked by the fact of war in 1939 and still fought it until 1945.[85] At the same time, British historians have marvelled over R. Graves, S. Sassoon, W. Owen, and others whose antiwar literature is full of admiration for heroic combat if it is properly fought and liberated from the loathsome ways of the British military establishment,[86] and over the remarkable transition from neutralist sentiments to all-out mobilization in the late 1930s.

It seems that there are two interrelated ways to explain this seemingly contradictory behavior of Europeans. First, war was tolerated and supported in defense of real and presumed national privilege. Even opponents of war found it difficult to oppose imperial "pacification" campaigns, and all except the most outspoken pacifists in Germany and Austria found it difficult to reconcile themselves with the rise of Poland, Czechoslovakia, or Yugoslavia. If there was a difference between the western nations and Germany, it consisted in the fact that the German sense of privilege was Eurocentric, whereas the British and the French ones were focused on the nonwestern world. This sense of privilege was commonly laced with the social construction of ethnic identity and superiority and it was always a defense of hegemony, be it in Europe or overseas. It is not by chance that this potent notion of privilege and hegemony condensed in the war against Bolshevism in 1941, which was not just popular in Germany but in all of Europe.[87] The inclusion of the mass of the population into a system of European privilege over others became a powerful source of national identity in which the struggle for mass participation was linked to supremacist dreams of social hegemony in an imperialist world, which was based on the destruction of all possibilities for autonomous development on the part of subordinated societies. The origins of these conditions can be traced back into the time before World War I, but they run like a red thread through the course of European history in the twentieth century. They are most clearly expressed in the simultaneous expansion of welfare services, consumption, and arms races in this period.[88]

If the defense of privilege was the main external thrust in the process of militarization, the rise of war and violence as class- and gender-defining metaphors—as means of establishing social and individual identities—was of equal importance. A watershed in European history was reached when, in the 1890s, war became an increasingly central aspect of bourgeois identity formation and when a new Right began to replace property and family with war as the main link between the (male) individual and the nation, and when maleness was defined in terms of guns and missiles rather than in terms of civil status.[89] This points to the rise of a common European and western popular war culture.

This development is closely tied to the rise of the corporate organization of European societies and to the subordination of autonomous individuals under the imperatives of large-scale institutional domination. Germany's course is indicative; for Germany possessed none of the "buffers" of imperial nations to harmonize social conflict and it pursued an extreme course of corporate self-transformation in order to participate in the lateral competition over the terms of European hegemony over the world. It faced massive resistance not just from the Left and the working class, but increasingly also from a populist Right, which turned to war as a means of establishing social and individual autonomy. The German mobilization for war in World War I was a first step in a long transition toward the habitualization of war—that is, a condition in which the preparation and the use of violence were no longer seen as exceptional or as deviations from the norms of civil society but became their embodiment. This development reached a high water mark in World War II and in the Cold War.

World War I added European-wide momentum to the process, when even those participants in war who came to reject war did not altogether negate war as social metaphor. During the 1920s the politics of the Left as well as the Right was infused with combat metaphors. War ascribed status to individuals and lent meaning to the "work" of those who participated in it.[90] The meaning of war changed drastically in this context. War was no longer guided by instrumental or utilitarian reasoning—a calculated use of force for specific ends. Rather war became a strategy of identification, a symbolic and value-oriented form of social conduct.[91] From here it is indeed only a small step to a new kind of ideological radicalism: the ideal of a society that not only organized itself around the central metaphor of war but defined itself on the basis of violent exploration—a system of state and social terrorism that negated old social identities.[92] The fascists turned this discourse into practice and attempted to create a new society that habitually lived off war—a parasitic society. National Socialism was unique in practicing this ideology, but it developed within a common European war culture that outlived the defeat of the Third Reich.

I have stressed throughout this chapter the very different trajectories of the process of European militarization that came into its own in World War I. I have tried to show how social arrangements shaped this process. But it is equally important to point to the common signature of the militarization of Europe. It is the defense of national privilege amidst a growing impoverishment of the autonomy of individuals and their social existence. The militarization of Europe was the answer to the collapse of the nineteenth-century bourgeois order.

PART THREE
Militarization in Contemporary Europe and America

Chapter 5
The Militarization of Europe, 1945–1986
GORDON CRAIG

IN APRIL 1986, Theo Sommer, the senior editor of the Hamburg weekly paper *Die Zeit,* wrote an article arguing against West German involvement in President Reagan's Star Wars project in which he said: "Nowhere in the world are so many soldiers, so much war material, and so many nuclear weapons concentrated in such a compressed area as in the two German states. If it should ever come to a conflict between East and West, not much of Germany would survive."[1]

This is a striking statement, most of all because of its unconscious irony. If one thinks back to the year 1945, one remembers that it was the universal resolve of the victorious powers that Germans should never again be permitted to raise armed forces of any kind, a feeling that most Germans shared, and yet the combined armaments of the two Germanies today beggars anything that existed at the outset of Hitler's war in 1939. In 1945, moreover, it was the hope of all western governments that their own countrymen would be freed from the military burdens of the prewar and war years. Lord Bullock's biography of Ernest Bevin as foreign secretary makes vividly clear how intent the Labour government was upon preventing military expenditures from crippling its projected program of economic and social reform and what a difficult task Bevin had in persuading the Treasury and the rest of the cabinet that Britain's overseas commitments would require greater military outlays than they wished to contemplate.[2] Nothing was further from the British mind than that they might have to make a major contribution to a European build-up. This was true also of the French, who were resigned to military spending for the defense of their colonial holdings but had no plans for a large European force, for which there was no obvious need.[3] If they had been told that within ten years Europe would be divided into two military alliances and that they and the Germans would be partners in one of them, and that within forty years their alliance would have 2.6 million men under arms and more than eighty divisions deployed, armed with thirteen thousand tanks, twelve thousand antitank weapons launchers, six thousand tactical nuclear warheads, three thousand delivery vehicles, including since 1984 Pershings and ground-launched cruise missiles, and thirty-five hundred tactical aircraft, with supporting naval forces and strategical nuclear systems, and that this panoply of arms would be equalled and in some cases surpassed by the opposing alliance, they would have rejected this as the purest fantasy.

The origins of this militarization, or remilitarization, of Europe are to be found not in the economic desires of munitions makers or the ambitions of political generals (the favorite villains in interwar literature on the arms race) but rather, as Michael Geyer has written recently, in the domestic crisis that confronted all European governments after the war, the possibility that fragile national systems of authority would collapse, and the porous state of internal and external order. The prevalence and the critical nature of the problem of *Regierungsfähigkeit* or the ability to govern led to the internationalization of the search for stability, that is, to the intervention of hegemonial powers who were determined to establish order in their own interest.[4] The road to militarization began with this intervention and, more specifically, with the attempt to solve the most intractable of all postwar problems, the reorganization of the disarmed and occupied Germany. For the economic recovery of Western Europe, the integration of Germany into the West seemed essential, and when the Soviet Union showed no disposition to tolerate any German settlement that would lead to that end, the western powers led by the United States opted for a divided Germany. Out of the crises that accompanied western progress and Soviet resistance to that end came such successive actualizations of the militarization process as the Brussels Pact, the Western European Union, NATO and its counterpart the Warsaw Pact, and, eventually, the rearmament of both parts of Germany.

Once the process of militarization was under way, both the soldiers and the military-industrial establishment began to play a significant role. If French politicians resisted the idea of German rearmament, French military leaders were convinced as early as 1949 that it was necessary and they were relieved when the French Assembly's defeat of the EDC in 1954 opened the way to a speedier and less complicated satisfaction of their desires. Former German generals, meanwhile—Hans Speidel, Hermann Foertsch, and Adolf Heusinger—had persuaded Konrad Adenauer to drop his original idea of raising a federal police force to deal with threats to the public order or border incursions from the Eastern zone and to adopt a plan, which was strongly supported by the French military and the U. S. Department of Defense, for a regular army of twelve armored divisions with supporting naval and aviation units.[5]

The gratification that professional soldiers found in these expanded opportunities for the employment of their skills was matched by that of military-oriented civilian enterprises and occupations. There were many beneficiaries of the process of militarization, particularly after it entered its nuclear phase, and Richard Barnet is giving us only a partial list of these when he writes:

> The war economy provides comfortable niches for tens of thousands of bureaucrats in and out of military uniform who go to the office every day to build nuclear weapons or to plan nuclear war; millions of workers whose jobs depend upon the system of nuclear terrorism; scientists and engineers hired to look for that final "technological breakthrough" that can provide total security; contractors unwilling to give up easy profits; warrior intellectuals who sell threats and bless wars.[6]

On the other hand, there was in the early years strong opposition in the western countries to what was judged to be a dangerously regressive movement. No one could travel through Germany in the months following the conclusion of the Paris agreements of 1954 without being impressed by the scope and intensity of antimilitarism. In neither motivation nor expression was this movement primarily a political one. The Social Democratic Party was solidly opposed to the raising of the army that had been authorized at Paris, but so were the trade unions, which were not exclusively socialist; so was an important part of the Evangelical Church; so, apparently, was the bulk of that section of German youth eligible for military service, who showed their displeasure at the prospect by howling down government speakers at a meeting in Cologne and stoning the defense minister with beer mugs at another in Augsburg. And so—if we can draw any conclusions from the tremendous success of the antiwar novels and films of writers like Helmut Kirst and Carl Zuckmayer, which portrayed the military caste and army life in the worst possible light—were large sections of the general population, whatever their private political views. The hostility to rearmament stemmed from a complex of fears and resentments, but the common denominator was the memory of Hitler's war and the belief that possession of an army would be the first step in a sequence that would lead inevitably to the same disastrous end. These fears were not unknown in other countries.[7]

Indeed, they tended to grow quickly when as a result of the inability or reluctance of its members to meet the force levels set at the Lisbon meeting of the NATO Council in 1952, that council in December 1954 decided that NATO's future strategic plans would be based upon the assumption that atomic weapons would be used in defending the West from attack. Indeed, Field Marshal Montgomery soon announced that this meant not that such weapons might be used, but that they would be, a statement that elicited from Britain's leading military critic, Sir Basil Liddell Hart, two long articles in which he argued that this was a confession of inability to adjust defense to the nature of the attack and would unleash all-out atomic war.[8]

To West Europeans, who lived in a confined and highly populated area, the prospect of the use of tactical nuclear weapons was daunting, and understandably so. In June 1955, NATO held air maneuvers that engaged three thousand planes from eleven member states and were designed to test offensive and defensive capabilities in an atomic attack by a tactical air force against military targets and their supply sources and communications. The CARTE

BLANCHE exercise was held over an area comprising the Low Countries, part of France, and the whole area of the Federal Republic of Germany. When it was over, it was calculated that if the 335 bombs dropped during it in the area between Hamburg and Munich had been real, 1,700,000 Germans would have been killed and 3,500,000 wounded, not counting possible casualties from radiation.[9]

The release of these frightening figures greatly complicated Konrad Adenauer's problem of persuading the Bundestag to pass his Army Volunteers Bill. It also helped to create a European sense of apprehension upon which Radio Moscow played effectively during the Suez Crisis of 1956, when it made open threats of bombing London and Paris. And this in turn probably encouraged the mood in England that gave birth to CND, the Campaign for Nuclear Disarmament. From the middle fifties onward, various groups in England had been campaigning against nuclear weapons. These were brought effectively into a coalition by Canon Collins and Kingsley Martin, the editor of *The New Statesman;* and in November 1957 that paper printed two remarkable articles by the philosopher Bertrand Russell and the novelist J. B. Priestley. Priestley called for "a declaration to the world that after a certain date one Power able to engage in nuclear warfare will reject the evil thing forever." This appeal for unilateral nuclear disarmament elicited a strong response that threatened for a time to capture the Labour Party, although it was always, as Denis Healey said later, less a political movement than "a movement of moral protest against what it saw . . . as the obscenity of nuclear weapons," and as such it appealed to people of every party, creed, and occupation, as was clear in the gigantic meeting in Trafalgar Square at the end of the Aldermaston March of 1959.[10]

Even so, the fears and the agitations of the 1950s did not slow the pace of remilitarization. Adenauer got his Volunteers Bill through the Bundestag, and West German divisions entered the NATO battle line, a fact that led to a corresponding acceleration in the growth of Warsaw Pact forces. The debate over NATO strategy that had been caused by CARTE BLANCHE blew itself out, and NATO continued to rely upon an atomic rather than wholly conventional defense. CND did not capture the Labour Party and quickly dwindled in strength; and, in any case, the Conservative victory in the parliamentary elections of 1959 assured that there would be no change in the military or alliance policy. Instead—and this was remarkable, considering the passionate nature of the nuclear debate in the 50s—a mood of relaxed and unreflecting acceptance settled over the European scene that was to last for twenty years, during which no serious public effort was made to question the tendencies of military policy, the need for the enormous armories being created, or the uses to which they could be put.

For this acquiescent mood there were several reasons. Until the end of the 60s, the state of the world was sufficiently unsettled to make a strong NATO imperative, while the high cost of conventional as opposed to nuclear weapons discouraged fundamental changes in strategy and encouraged the gradual build-up of theater nuclear forces. At the same time, the general belief in American superiority in strategical nuclear capacity gave credibility to the deterrence theory that underlay western policy. Indeed, when *détente* set in at the end of the 60s and it seemed possible that a *modus vivendi* between the United States and the Soviet Union and a real measure of *Entspannung* (relaxation of tensions) in Central Europe were possible, it looked as if NATO's strategy had been successful in opening a way to a measure of control over the process of militarization. At the Moscow summit the first real measures were taken to regulate the competition in strategic weapons, and SALT I encouraged the hope of even tighter controls in that area, while at the same time galvanizing the Vienna talks on mutual and balanced force reductions in Europe.

Nothing came of these hopes. Long before the end of the 70s, Americans—perhaps for the reason that Henry Kissinger gives in the second volume of his memoirs, that is, their Manichean approach to foreign policy and their tendency to see every other power as either friend or foe, with nothing in between—had become disenchanted with *détente* and were telling themselves that the Russians had been using it as a screen behind which they cheated on all the engagements they had made at the Moscow and Washington summits.[11] SALT II was soon in deep trouble and did not survive the misguided Soviet invasion of Afghanistan. The prospects of any positive results from the Vienna talks on troop reductions in Europe shriveled and died in what rapidly began to seem a new ice age. From Washington, Europeans began to hear about windows of vulnerability that had to be closed by new military efforts, and the Republican candidate, who made no secret of his view of the evil nature of the Soviet adversary, began to call for the deployment of a new generation of nuclear weapons by the West. In an attempt to head this off, the West German Chancellor Helmut Schmidt conceived what became known in Europe as the NATO double decision, which stated that if the West were not successful within a period of five years in persuading the Soviet Union to resume arms control talks in a serious spirit, NATO would begin the systematic deployment of 108 Pershing missiles and 464 ground-launched cruise missiles in Britain and Western Europe, this deployment to be completed by 1988.[12]

It was this sequence of events that ended the twenty years of complacency and revived all of the anxieties and passions of the 50s. The American rejection of *détente* was bitterly disillusioning to Europeans, to whom the *Entspannung* of the early 70s had brought tangible gains in the form of trade and improved East-West communication and travel and who had begun to believe in the possibility of a return to a normal, crisis-free world. Many of them were unconvinced by American arguments about Soviet bad faith or by the Reagan-Weinberger attempts to prove by statistics and pictures that the military balance had shifted to the disadvantage of the West. Starting in the early 80s there was a decline of confidence in American leadership that was particularly strong among younger voters and university-educated professionals and intellectuals, and this was encouraged, as the time limit envisaged by the NATO double decision passed without productive arms talks, by the perception that the Reagan administration had not tried seriously to promote these talks because they were intent on their plans to modernize their European nuclear forces.

The peace movement of the 1950s now revived impressively. In England, there was a renewal of CND, now taking the form of demonstrations designed to block the stationing of the missiles there. A strong international peace movement spread through Scandinavia, the Low Countries, and West Germany, reaching its height in the latter country in the fall of 1983 with demonstrations, marches, and peace seminars at Krefeld, Bremerhaven, and Mutlangen that were strongly supported by the churches, the universities, the left wing of the SPD and the Greens. To this movement the German Democratic Republic proved not to be immune, for a *Friedensbewegung* (peace movement) grew within the Evangelical Church there, with some unprecedented public agitations; even Erich Honecker, appalled by the prospect of what new military outlays would do to the country's faltering economy, sounded on occasion like a timid peacenik.

At the same time, there was a renewal of the strategical debate of the 1950s. NATO plans were once more the subject of critical attention, and for the first time in twenty years there was serious discussion of alternatives to deterrence policy. Articles on neutralism began to appear in the public print. Debates were held on the possibility of creating a nuclear-free zone in Central Europe, on the efficacy of no-first-use pledges and a nuclear freeze.

Once more, however, as in the 1950s, the intense agitation failed to check the relentless progress of militarization. Many Europeans doubtless regarded NATO commander General Bernard W. Roger's remark that "we must arm in order to disarm" as an exercise in dubious logic, but in the end they accepted it.[13] Or they seemed to, for certainly the NATO governments fell into line and so did their parliaments, and despite the *Friedensbewegung* the deployment of the Pershings and the cruise missiles was approved and in 1984 got under way.

This was, I suppose, inevitable, because the bulk of European opinion, while opposed to the arms race, is nevertheless uncertain as to the means and ambivalent about the advisability of stopping it. We have a considerable amount of information about public opinion in these matters, but, while it is interesting on specific points, it seems to me to reveal the quality that the Germans call *Ratlosigkeit,* or perplexity. James Scaminaci of Stanford University, who has made a careful assessment of data collected during the past five years by means of Gallup, Harris, Emnid, Allensback, USIA and Eurobaromètre polls, has reached some tentative conclusions about the present state of European opinion that bear this out. He writes that the Soviet Union is still viewed as a threat to Western Europe, but that the intensity of this conviction has been dulled by *détente* and that, although few people seem to trust the Soviet Union, few believe as strongly as they once did that the Soviets will attack. In the second place, confidence in the United States as leader and protector of Western Europe has declined and concern over the wisdom and judgmental balance of American policy has increased. In the third place, the vast majority of Europeans still believe that NATO is essential to their security, and only a significant minority believe in neutrality, even in Holland, which is sometimes mistakenly considered to be the center of the neutralist movement. But the popularity of NATO has sharply declined, particularly among young people.

In the fourth place, while few Europeans support unilateral nuclear disarmament, most of them are unhappy about increasing nuclear arming. Their assent to the deployment of the Pershings and the cruise missiles was much more conditional than it appeared in administration readings in this country, and there is continuing opposition as it proceeds. Meanwhile, interest in nuclear freeze and nuclear-free zones remains high. In addition, Mr. Scaminaci has written, the contradictions revealed by the polls indicate that one of the salient effects of the introduction of nuclear weapons in Europe may have been to strengthen the determination never to use force again that was prevalent immediately after the two World Wars. One thing that seems certain is that Europe has lost its old glorification of war, military heroes, and the practice of power politics.[14] It is possible that they have carried this too far in certain countries. In West Germany, whose army has been one of the most effective components of the NATO battle line and that has despite the fears of many people in the years 1950–54 been a model of decorum in domestic politics, the Bundeswehr can hardly hold a public ceremony without inviting agitation, protest, and clashes between rioters and the police.[15]

What about the economic and psychological effects of the militarization that I have been describing? With respect to the former, it can be said that ever since the days when Ernest Bevin and Hugh Dalton fought their battles over the respective shares of public funds that were to be alloted to national security and public welfare, military expenditure has been for all European governments a burden and a restraint, and the ballooning costs of weaponry in the last years have greatly complicated the problem. The International Institute for Strategical Studies's annual volumes on *The Military Balance* are filled with references to weapons systems that are "behind schedule and much over cost" or sad items like "Portugal is faced with an extensive modernization program which is quite beyond its economic resources." In the volume for 1985–86, the editors comment on the nature of the continuing problem in a way that hints that European countries may have suffered, although to a lesser degree, the same kind of deprivations and cuts in domestic development programs that we have experienced in the United States in the last five years. They write:

> The economic recovery of the NATO Allies continues to lag behind that of the United States. Their economic performance has been characterized by rising unemployment, mixed economic growth, declining exchange rates—which have a significant effect upon a country's ability to buy military equipment from the United States—severe capital flight to the higher interest rates paid in North America, mixed success against domestic inflation, and social pressures on the budget. Virtually all West European states have, with varying degrees of success, introduced measures to control inflation and to cut their respective budget deficits, but must meet rather ambitious social welfare programmes and high unemployment levels which place heavy demands on welfare budgets. Nor is there the same perception of imminent

threat as in the U.S., although there is no intention of compromising freedom. In the light of these circumstances, most West European governments believe that they are spending all that they can afford on their defence establishments.[16]

Any assessment of psychological results of the militarization of Europe must be tentative because of the fragmentary nature of the evidence. That fear of war has become a part of European thinking goes almost without saying, and the salient fact in this regard is the sharp increase of apprehension in the 1980s. David Capitanchik and Richard Eichenberg have reported for Great Britain that in 1964 barely 40 percent of respondents to a Gallup poll were worried about the possibility of worldwide war in which nuclear weapons would be used, whereas in a similar poll in 1983 over 60 percent indicated that they feared such a war. In other polls, 38 percent of British respondents, 40 percent of French, and 32 percent of West Germans said in 1980 that they believed that a nuclear war would occur within ten years, an opinion concurred in by a meeting of scientists, military experts, and peace researchers at Groningen in Holland in 1981.[17]

The extent to which this anxiety can dominate normal life is described by one of Germany's leading dramatists, Botho Strauss, in his volume of essays, *Paare Passanten,* written in 1981. He wrote:

the cruise missiles, neutron bombs and other devices before my door . . . serve only as arguments for the thesis that highly civilized populations have the tendency to destroy themselves. The feeling "After all, everything we do is vainglory and worthy only of ridicule"—this originally religious sense of insignificance before the glory of God and the starry heavens comes to us now from this world, and very close. . . . The bomb has been there for a long time, installed and ready for use. Most of the time we forget it. Nobody can run around with this . . . incision in his head and remain unaffected. But the threat is total, ever present, never to be forgotten, but also unthinkable. . . . *Si vis pacem, para bellum.* If in the end you want peace, the peace of the graveyard,

> the peace of ruins, then arm for war. Deterrence is purchased at the price of limitless terror in everyone's breast, and perhaps also at the cost of our spiritual resources of resistance, of our economy, of our society.[18]

It is no accident that a decided apocalyptic strain has entered into German literature. The theme of the DDR novelist Christa Wolf's impressive book of 1981, *Kassandra,* was the total destruction of a society because its leaders would not heed the plain warnings that they had of the imminence of such destruction. The main character of Gunter Grass's new novel *Die Rätinn* is a female rat who persistently talks of the human race in the past tense, as victims of self-immolation, although the human protagonist feebly protests that this is not true, or at least not yet. This tendency fits into what Fritz J. Raddatz has described as a retreat in recent literature and theater from both history and contemporary reality into myth, citing the high incidence of plays, novels, ballets with classical themes, and indeed themes of fate and retribution—*Oedipus* and *The Oresteia,* Reinhold Hoffman's ballet "Dido and Aeneas," Rolf Hochhuth's drama *Judith,* Michel Tournier's novel about National Socialism, *Der Erlkonig.*[19]

Would it be entirely unreasonable to conclude that these tendencies reflect a disturbing national psychological mood? Saul Friedländer tells a story about a twenty-five-year-old German who dreamed he was alone in a house after what he sensed was a nuclear disaster and knew that he should not go outside into the radiation-poisoned streets but was drawn into them against his will and walked through the dead city toward a large set of buildings that he knew he was forbidden to approach, a complex that had ramps and loading areas, all of which, he knew, had something to do with gold, but what he could not discover because he woke up. This, Friedländer explained, was not a nuclear dream, as the young man's psychiatrist concluded, but a dream about the Holocaust, for the buildings with the loading areas obviously belonged to a concentration camp, and gold is a well-known symbol for the Jews. This is persuasive enough, but it was of course also a dream induced by the fear of nuclear war, or perhaps a dream of two holocausts, one in the past and one in the future, and in this sense a dream of retribution. And in any case, it was not a healthy dream.

It is perhaps not exaggerated to see other disturbing results of militarization in German society. German's leading writer of children's books, Michael Ende, in a recent radio broadcast in Bremen, described the atom bomb as the culmination of mankind's three-hundred-year search to uncover the secrets of the universe, the final product of human rationality, and this proves, he said, that human beings must for their own self-protection create a counterweight that, by its very nature, would be inexplicable—a secret, a *Mysterium,* into which human beings must plunge with body and soul in order to understand in a quite different way, not in a rational way. No one who remembers German history can feel very comfortable at this resurfacing of Romantic pleas for irrationalism, but then no one who considers the present state of European armaments will find it easy to believe in the rationality of our times.

Peter Iden of the *Frankfurter Rundschau* has said that West Germans suffered from a sense of the impermanence of their existence that was reflected in the catch-phrases that they used in their ordinary conversation. To paraphrase him, people "were constantly using words like *wahnsinnig* (crazy) and *geheimnisvoll* (mysterious) to describe quite normal and ordinary things, as if to express surprise over their existence or ability to function. They were always saying '*Alles klar!*' as if desperately resolved to give clarity to a situation that no longer warranted it, or assuring others that such and such a thing would be done '*mit Sicherhheit*' (most certainly) so emphatically as to suggest that they no longer believed in any form of durability."[20]

This air of tentativeness has, if anything, increased and it is not confined to Germany. The declining belief in the durability of peace is having profound effects on personal behavior in all

European countries, as can be seen in such things as, on the one hand, a growing fatalism, a frustrated distrust of political leaders, and a withdrawal from political participation, a new focus on regional and environmental problems, and an internalization of life at the expense of the *polis,* and, on the other hand, in participation in grass-roots, direct-action movements that demand immediate and total solutions for complicated political and military problems.

Chapter 6
Beyond Steve Canyon and Rambo: Feminist Histories of Militarized Masculinity

CYNTHIA ENLOE

MILITARIZATION IS a societal process, just as urbanization or industrialization are societal processes. Gendering, the masculinizing and feminizing of certain roles and symbols, has been as central to militarization as it has been to other social processes. Yet the relationship between gender and militarization remains obscure, in part because we assume that the military has always been men's business, always requiring a certain type of masculinity.

Now that large numbers of persons of the female sex are also in the armed forces, it would seem that gender would be a subject of considerable interest. However, recognition of gender is not just a matter of giving women a place in military history, for gender was there long before women were. Gender is what is understood as the differences between the sexes, what is ascribed to masculinity and femininity, what men and women are supposed to stand for, whatever they may actually be. The gendering of the military thus has a long history predating the modern female soldier; and the militarization of gender affects men and women who have nothing to do with the military as such. As societal processes, gendering and militarization are inseparable, though their relationship has varied widely over time and space and is never unproblematic or uncontested. The definition of sexual difference has been constantly changing in the present century. The idea of what militarized masculinity should be was not necessarily the same in the First and Second World Wars as it has been in the post-1960s wars. Nor does the militarization of either masculinity or femininity necessarily take the same form or depend on the same lures and sanctions to be sustained in Japan as it does in Britain or in Vietnam. How "motherhood," "sweetheart," "buddy," or "nerd" as proscriptive characterizations will support or undermine militarization has to be explored in the context of a given time and place. This essay will only suggest some of the dimensions of the gendering process on which the militarization process has depended in order to make it both more visible and more problematic.

I grew up with a blue-haired Superman fighting Nazi spies. Golden-haired Steve Canyon was my comic strip passport to wartime Asia. My father, serving as a physician with Wingate's commandos in Burma, was part of the group of British, American, and Australian soldiers who inspired Milton Caniff to write his famed Steve Canyon strip. In a process analogous to novelist Bobbie Ann Mason's teenage Kentucky girl making sense of her father's death in the Vietnam War by watching tv reruns of M*A*S*H, I began to formulate ideas about what it was like to be my father in World War II Burma by reading Steve Canyon comics.[1] Perhaps even the men who were in those unconventional commando units mingle their own memories today with images from Steve Canyon and Terry and the Pirates. My mother, on the other hand, didn't inspire wartime comic book creators. She was neither the Dragon Lady nor Cheetah. And I don't remember her ever reading these strips. She took care of my brother David and me as essentially a single parent for months at a time. She cooked meals and provided a home-away-from-home for my father's army air corps friends when they and he were back in the United States between overseas duties.

Each of us in our own way took for granted a symmetry between masculinity and militarism, whether that was presented in popular culture or in the lives of family members. I didn't give any thought to whether the link between masculinity and militarism had to be forged. I didn't wonder about how that forging was done. I didn't imagine debates, obstacles, manipulative strategies, setbacks, or costs.

It is only now, forty years later, reading my mother's enticingly cryptic diaries from those years, listening to my father describe plans for commando reunions, comparing Steve Canyon reprints with the latest Rambo toy narratives that I have begun to ask questions.[2] How have men from different cultures had their notions of manhood—and womanhood—shaped and reshaped by officials (largely male) so as to permit governments to wage the sorts of wars they have imagined to be necessary? What contradictions or failures have had to be camouflaged in order to allow the ideological symmetry between masculinity and militarism to appear unproblematic, "natural?"

In the late 1980s, Rambo has caught the imaginations of millions of people. For a time it was the top film in both the United States and Britain. It is being viewed on VCRs from the mountains of Luzon in the Philippines to downtown Helsinki by people as different as Filipino guerrillas and anti–Cold War Finns. Sylvester Stallone's character has gone on to inspire fashions, dolls, and television series. "Rambo" has slipped quickly into the global lingo of adults as well as children, of militarism's critics as well as its enthusiasts. For many of us, "Rambo" has become, I think, a handy shorthand for a complex package of ideas and processes that we believe are dangerous to all women and many men. Rambo's brand of militarized masculinity is being compared and contrasted by students of popular culture with those of World War II movie idol John Wayne and contemporary National Security Council bureaucratic entrepreneur Lt. Col. Oliver North.[3]

One of the most important contributions feminists have made to the analysis of war and peace—and militarized peace—is their descriptions of how notions about masculinity and femininity have helped to promote and sustain the military. While other critical analysts have given economic, racist, and bureaucratic patterns their prime attention, feminists have concentrated on the social constructions of gender. The accumulation of more and more evidence from more and more societies has made feminists increasingly confident in asserting that the omission of gender—femininity and masculinity—from any explanation of how militarization occurs not only risks a flawed political analysis; it risks, too, perpetually unsuccessful efforts to roll back militarization.[4]

At this juncture in the historical evolution of feminism, "Rambo" has appeared. Its remarkable cross-cultural popular appeal—an appeal fueled by immense infusions of corporate capital—has seemed to confirm the feminist analysis. First, social constructions of masculinity—not just elite interests or state bureaucracies and their cosmologies—are serving to entrench and extend the grip of militarism. Second, militarism's reliance on particular forms of masculinity apparently exists in societies with otherwise different cultures and at different levels of industrialization. Third, militarizing masculinity cannot succeed without women also being made to play their parts in the militarizing process; although vital, those parts must be kept ideologically marginal.[5]

I find these explanatory arguments persuasive. But I'm beginning to wonder whether they are enough. Specifically, I think we may need to test two new sets of hypotheses in our continuing exploration of how militarization works:

(1) All societies use, though each in its own way, ideas about masculinity and about femininity to organize themselves for state-controlled violence.

(2) It requires more than just one form of masculinity and more than one form of femininity to make militarization work in each setting.

In the following pages I will consider these hypotheses as they help reveal how militarization occurs in: (1) coping with the experience of wars past (won or lost); (2) the militarization of the Third World; (3) the internationalization of the military; (4) the militarization of sexuality; (5) the militarization of a society's civilian sector.

Rambo is not Steve Canyon. Steve Canyon may get a bit confused when he moves from comic strip World War II to comic strip war in 1950s Indochina. But he remains a character on a winning side. While somewhat of a maverick, he doesn't feel he's at war with his superiors. World War II, in this simplistic portrayal, seemed to involve a militarization of masculinity that served the state without depriving the white American male of his sense of individuality and his emotional attachment to women. The Rambo character is quite different. He openly defies his superiors. He tries to reopen a war that his state authorities want declared "over," if not won. He is so unconnected to either his fellow men or women that he rarely speaks in whole sentences. Rambo is a peculiarly "post-Vietnam" type of American militarized male. His message for men is about how to cope with national humiliation and elite betrayal: by resorting to individualistic military adventurism that defies official hierarchies but restores a nation's "pride" in its military.

But do men in other societies have the same responses to national humiliation? How have Belgian, Dutch, German, Chinese, Polish, Egyptian, Italian, Japanese, or French constructions of masculinity been affected by twentieth-century military losses? "Humiliation" is both a gendered and an enculturated emotion. So we might expect it to be militarized in quite dissimilar ways in different countries, with the consequence that "Rambo" will be absorbed (or rejected) in quite dissimilar ways.[6]

The very definition of a war "lost" is problematic. As Finnish feminist Eva Isaksson asked, "Did you ever consider whether men in countries that lost their wars really think that they actually lost?"[7] It may be that men and women in the same country ("on the same side") carry into contemporary political action quite contrary presumptions about whether there is anything to feel humiliated or repentant or defiant about. Moreover, men of different social classes may conceive of a war's outcome in ways so various that it produces quite dissimilar relationships between their senses of masculinity and its relationship to the state military.

Thus it is always useful to examine the current trend in militarizing masculinity in the context of what particular men imagine to be "the last war." For example, Klaus Theweleit, the German historian, has delved into the most intimate fantasies of men in the Freikorps. He argues that this particular group of men were militarized in large part by their desperate flight from the feminine. But this flight and the hatred for women it produced, which was so effectively manipulated by conservative authorities from 1918 through World War II, wasn't fueled by an ahistorical misogyny. According to Theweleit, it was rooted in these men's particular experience of a war lost, the First World War.[8] In other words, to understand how the men in particular classes, ethnic groups, and nations adopt or reject notions of military humiliation and redemption we need to take seriously the militarization dynamics of "postwar" eras.

Let us take another example, one that is not idle speculation in 1989. How does military humiliation resonate among women and men in contemporary Japan? Despite U.S. government pressure on the Japanese to remilitarize and despite Prime Minister Nakasone's apparent desire to play down Japan's World War II errors and to revive the nation's military strength, Japanese popular notions of masculinity to date do not seem easily remilitarized. Yet Japanese feminists are closely monitoring what they believe are important, if subtle, efforts by the government to transform the postwar economistic model of Japanese masculinity in ways that will make it more amenable to a U.S.-backed military buildup.[9]

Do men in other societies sift an alleged national humiliation through a militarized sieve in ways that make them respond to "Rambo" in precisely the same individualistically defiant way that American men seem to? I doubt it. In every country where "Rambo" seems to have become a cultural hero, students of militarization should look to see exactly what it is in the character and the narrative that is attractive to men. What do those tendencies, in turn, tell us about how men in those countries might be drawn to support militarism or reject it?

There is a second territory for our investigation of possible varieties of militarized masculinity: cross-national military training programs. Is "a drill sergeant a drill sergeant a drill sergeant"? The British government is an old hand at militarization. One of its successful empire-building strategies was to build armies out of local colonialized labor. To do that, British officials had to find ways to persuade local male rulers that their personal authority and status would be enhanced if they would allow the British to build them "proper" armies filled with "proper" soldiers. Then they had to persuade thousands of male peasants and nomadic herdsmen that their manhood would be enhanced within their own communities if they would enlist in the newly created British-controlled armies. There was nothing automatic about either of these processes. Creating modern armies out of traditional materials was never an easy task. It involved a subtle mix of persuasion and coercion, always varied by time and by place.[10]

This complex historical process was more than a matter of exporting British notions of militarized manhood to different ethnic communities in India, Nigeria, and Malaysia. Reading old training manuals and eavesdropping on British colonial officials' reminiscences suggest that considerable adaptation had to go on to make this imperial strategy work.[11] That is, masculinity could not be militarized in Scotland in exactly the same ways that it could be militarized in India or Nigeria. Furthermore, women in each of these countries—Britain, India, and Nigeria—were thought of as playing slightly different roles in order to sustain the "manly soldier" needed by the empire. This was rarely talked about in formal reports and typically is ignored entirely by military historians. But it was analyzed at length by feminist reformers in the 1880s, when, having won the repeal of the patriarchal Contagious Diseases Acts, they launched an international campaign to expose the British government's policies toward military prostitution in India. Campaigners writing in their journal *The Dawn* noted that British male officials believed that, for some reason, Indian male soldiers didn't use Indian women as prostitutes in the same ways as British male soldiers posted in India did. The bureaucratic debates over how to reduce venereal disease among British soldiers stationed abroad became a discussion of differences between British soldiers' and the empire's foreign soldiers' militarized sexuality.[12]

World Wars I and II both were fought by major powers with men (and, less visibly, with women) from their respective colonies. The British government used Indian and Caribbean men as soldiers; the Japanese government used Korean men; the French used Vietnamese and North African men; the Germans used East African men; the Americans used Native American, Hawaiian, Puerto Rican, and Filipino men. Out of these wars male officials drew lessons about what sorts of masculinity "worked" in combined military operations. But, with few exceptions, we know little about how these men from the colonized societies experienced militarized standards of manhood or how such experiences shaped postwar nationalist movements and the relations between local men and women and imperial and colonized men. Nor have we been curious enough about how postcolonial military policies concerning training, leadership, sexuality, marriage, or weaponry have been molded in part by these World War I or II interactions between colonizing and colonized military men. Studies of the career patterns and military curricula of South Korea, the Philippines or Jamaica might be one place to start in understanding the varieties of military cultures around the world.

Training is rarely devoid of historically gendered "lessons." The British government, despite a shrunken empire, continues to energetically export its military training expertise. It is often an instrument for promoting the sale of military equipment. But it is more than that. In November 1986, the British government signed an agreement to start training Mozambiquan soldiers. Reports suggested that it was the Zimbabwean regime of Robert Mugabe, a supporter of the besieged Mozambiquan FRELIMO-led government, which smoothed the way (the western press called it "acting as the midwife") for this military training agreement. Evidently, Zimbabwean officials have been pleased with the British military programs impact upon its own military, a force that had to be rebuilt at the end of the Zimbabwean revolutionary war. Does this mean that contemporary British military strategies for turning 1980s British middle-class men into officers and working-class (often in fact unemployed) urban men into useful soldiers can be applied to rural Zimbabwean men of different ethnic groups with no adjustments?

What do officials of the importing governments hope to gain by subjecting their military men to foreign training? Perhaps such policy choices imply an elite's disappointment with their society's current constructions of masculinity. Perhaps the acceptance of British or United States—or French, Cuban, East German, or Israeli—military training teams suggests that a country's male officials imagine their own male citizens to be "undisciplined." That is, maybe they believe that the "traditional" construction of masculinity in their own societies (or at least in the ethnic communities from which they choose to recruit most of their soldiery) is too "wild" or "disorderly" to serve the regime's own goals of national unification, social order, and state security. Alternatively—or perhaps simultaneously—the present elite might imagine that the conventions of manhood among their male population are too dismissive of the sorts of modern weapons technology that requires literacy and patience. Thus they have concluded that American or British or East German men's approaches to militarizing masculinity will produce the kinds of soldiers they think they need. And the exporting officials are more than eager to comply.

Elites who invite foreign soldiers to train their men may go so far as to imagine that their own men, subjected to such manipulation, will return to civilian society after their military service as more productive male farmers, more loyal government supporters, more responsible fathers. Certainly all of these assumptions are encouraged in the exporting governments' own recruiting promotions. What do these same officials assume, then, about their women citizens' reactions to their fathers', husbands', and brothers' new attitudes?

The flow of consequences from the internalizations of military training has never been one-way, however. The men who have acted as colonial officers, international liaisons, or foreign advisors have returned home from their assignments with lasting notions about which masculine traits in their own societies make them, as men, "naturally" better, braver, more inventive, more professional, more disciplined—and better soldiers than the men from the other cultures they have been sent to train. Usually there is a strong element of ethnocentrism or outright racism mixed into these paternalistic militarized, masculinized memories.

In the January 1987 issue of an American magazine called *Vietnam Combat,* a publication devoted to recalling the brave deeds of rank-and-file American men (not women) who served in Vietnam, there is a full-page photo showing a tall, white American soldier wading through a Vietnamese rice paddy carrying a toylike M–16 rifle. Next to him walks a shorter Vietnamese man in battle fatigues; he is carrying a much more potent and heavier piece of weaponry. American advisors referred to their Vietnamese advisees as "indigs." What ideas did this American militarized Big Brother bring back to the United States about his militarized little brother's bravery, loyalty, intelligence, sexuality? We won't have a true picture of the complex political experience we so glibly call "the Vietnam War" until we have a feminist history of the ARVN. And whatever the masculinized, militarized memories American military advisors brought back with them, they are likely to have institutional roots back to 1950s training programs in the Philippines and tentacles reaching forward to 1980s training programs in El Salvador and Honduras.

In other words, since the nineteenth century, the militarizations of masculinity have been motored by both domestic and international processes. And these historical and contemporary processes can be researched and documented. We are not dealing here with amorphous cultural ebbs and flows. There have been and are institutions to be studied: their routines, their policies, their debates, their language. The U.S. Army's Fort Bragg, Britain's jungle schools in Malaysia and Nigeria, the Philippines Military Academy, the U.S. military's international training school in Panama—each is a potential site for researching the international processes of militarizing masculinity.

There is a third arena in which masculinity is militarized: joint military maneuvers. These institutions are a post–World War II phenomenon, based largely on elite male officials' presumptions about what worked and what didn't work during World War II military operations. Milton Caniff's wartime comics hint at the dynamics between men from different cultures being thrown together by their governments in joint operations. Usually he has his Australian, Scottish, and American characters joke with one another about their respective manliness. The setting, typically, is friendly competition with one another for local women.

As the major powers have grown more determined in the past four decades to create separate and competing networks of military alliances around the world, more women and men are being pressed to coordinate their resources, skills, and fears so as to guarantee centralized strategic military actions. Most critics of these sprawling global alliances focus on the friction generated in trying to synchronize disparate legal and industrial conditions. For instance, in recent years there have been rumbles of discontent throughout NATO: the American partner's call for the "standardization" of NATO tanks, rifles, and computers seems to be one more attempt by Washington to compel everyone else to "Buy American," to the disadvantage of their own needy defense contractors. Within the Warsaw Pact, similarly, there are simmering discontents over the Moscow-imposed divisions of labor, rationalized by references to military alliance cohesiveness.

Much less talked about inside these alliances is how incompatible forms of militarized femininity and masculinity are reconciled. Most obvious are alliance members' differing policies toward allowing women to serve in the uniformed services. These policies are shaped and reshaped by legislators and bureaucrats concerned in no small measure with protecting the ideological bond between masculinity and military service. Within NATO, in 1986, the U.S. military is 10 percent women, while its Italian and Spanish allies still prohibit women from serving. Within NATO there also are different policies toward women in combat: the Dutch now allow women to serve in jobs defined as "combat," while the British take a firm stand against combat assignments for women and the Americans are on the brink of another public legislative debate over both the definition of "combat" and the military rationale for excluding women from such jobs. In fact, as each male elite tries to negotiate its way between gender conventions and military "manpower" needs, each is developing a different definition of what exactly "combat" is. As in World War II, "combat" is treated like a sensitive ideological instrument: it has to be defined narrowly enough so that military planners can recruit and deploy women to fill the perceived gaps in manpower; yet it must be defined broadly enough to preserve what are imagined to be the militarily useful distinctions between men and women. Sometimes defining "combat" has been a political high wire act.[13]

There also are growing differences among NATO's militaries concerning the alleged relationship between homosexuality and "national security." The Canadian parliament in 1986 was pressed by feminists to drop the exclusion of lesbians and gay men from the Canadian armed forces. In the United States, political mobilization among lesbians and gay men is light years ahead of where it stood in the 1940s. One consequence is that the U.S. government today faces multiple court actions brought by women's and civil liberties groups challenging the legality—and the logic—of discharging women and men accused of homosexuality.

In virtually every NATO military the constructions of masculinity and femininity—and the relationship of each to military preparedness and national security—are in flux. The Dutch, West German, U.S., French, and Italian governments—as dissimilar as they are in terms of historical memory and in contemporary strategic roles—in the late 1980s are all carrying on discussions about how to make more military use of "their" women: they are fretting about women's decisions to have fewer children, but they refuse fundamentally to alter their notions of military "manpower" needs. How each regime goes about resolving this gendered strategic dilemma will vary. Among the critical variables will be: 1) the strength of each country's women's movement and lesbian-gay movement; 2) the historic ties among civilian manhood, the conception of citizenship, and the state's military; 3) the current elite's definition of "national security"; 4) the availability of young men from the classes and ethnic groups that the military commanders trust. The gendered dynamics of an multinational military alliance such as NATO will be the product of each of these domestic relationships worked out through the unequal structure of NATO itself.[14]

What would the Warsaw Pact's internal politics look like through similar lenses? Ethnic Russian men at the top of the Soviet military command structure worry more about their growing dependence on Asian Soviet men for rank-and-file soldiery. Historically, this seems to be a new worry. It derives not from the lessons these men have drawn from Soviet experiences in World War II, but from the meanings they are assigning to the Red Army's experiences in Afghanistan. Will they move toward using more women, especially "European" Soviet women, in uniformed military jobs? And if they do (there is evidence that this is already happening), and if they find they can control the dynamics of masculinity and femininity sufficiently not to damage their military performance, will they then encourage other Warsaw Pact regimes to recruit more women as a complement to and not a substitute for continued pressure on women from trusted ethnic groups to produce more future soldiers? Already Rumania and Yugoslavia make increased use of women in their national security formulas—especially in local militias and in the name of "home defense." This institutionalized rationale avoids fundamentally challenging existing ideological ties between manhood and militarism. But both of these regimes are outside the core of the Warsaw Pact, making their gendered military innovations less likely to affect intraalliance dynamics.[15]

Joint maneuvers are feats of logistical and political management. They are also tests of militarized gender management. For the success or failure of any joint maneuver hangs on whether soldiers and officers of different cultures can learn to trust each other. And trust depends on some shared understandings of "bravery," "loyalty," "reasonableness," "reliability," and "skill." Each of these concepts is gendered. That is, each is infused with presumptions about proper behavior for "real men" and "real women." But are they each gendered in precisely the same way in each society whose soldiers are thrown together on a supposedly hostile battlefield? If they aren't, how do sergeants, majors, and generals try to reconcile those differences so that the joint maneuver will "work"?

It may be that different militarized masculinities are not fully reconciled; they only are band-aided over with misogyny. How often do Honduran and American men on "Big Pine" joint maneuvers camouflage their basic distrust of one another by trading jokes about the Honduran women working as prostitutes around their camps?[16]

Military medicine is a fourth site for making visible the historical processes involved in militarizing masculinity. In 1909, U.S. Army physicians conservatively estimated that two hundred men out of every one thousand troops were being hospitalized for venereal disease. According to historian Allan Brandt, between April 1917 and December 1919 the U.S. military as a whole recorded 383,706 male soldiers having been diagnosed with either syphillis, gonorrhea, or chancroid. By the end of World War I these numbers had set off a debate inside the American military establishment over which conceptualization of militarized masculinity best served the U.S. war effort in Europe.

Allan Brandt describes how Progressive reformers pressed the military to rely on the "cleansing influence of war" to restore the moral purity of American manhood. Following this policy, military officers would appeal to American male soldiers' sense of their own moral aspirations and to their capacities for self-control.

But many military authorities were skeptical. Their notions of manhood in khaki were less optimistic. Instead of trusting the moral self-control of their troops, these commanders argued that they should be allowed to institute compulsory medical exams for soldiers and, much more controversial, that they should be authorized to distribute chemical prophylaxes to their men. They based their policy stance on the assumption that "men will be men," especially when taken away from the restraining influences of the feminized home environment. These military men cared less about promoting war's "cleansing influence" than about ensuring military victory.[17]

Every military strategy for preventing sexually transmitted diseases among male soldiers has had as its companion a strategy for controlling women, local women and foreign women. One can read the political history of venereal disease and now AIDS as an account of how military officials have sought to control women and the idea of femininity for the sake of getting and keeping the kinds of militarized men they have wanted. Indoctrination and basic training courses for male soldiers have had to be carefully designed; those for women soldiers have had to be fashioned quite differently. Guidelines for the practice of military medicine and public health officials have had to be assessed with an eye to manpower and morale needs. Health department officials operating in the towns near military bases have had to be persuaded to conduct their business in ways that support the kinds of masculine and feminine behavior thought to enhance the military. Training courses and policies for "doughnut dollies" and other women volunteers have had to be fashioned so that those women's work bolsters rather than shakes militarized men's morale.

Civilian and uniformed policymakers have not found it easy to design policies that maintain the military's legitimacy in the eyes of the wider citizenry, with its own expectations about appropriate masculine and feminine behavior, while simultaneously ensuring a kind of soldiery presumed to optimize military effectiveness. Policymakers have worried over compulsory genital exams for their male soldiers, fearing they might jeopardize the men's morale.

They have debated whether women in the military should be given the same access to contraceptives as men in the military. There have been twists and turns in military policy regarding whether to allow prostitutes on military bases and whether to leave the physical exams of suspected prostitutes to local civilian authorities. Policymakers have seemed to be in a perpetual state of uncertainty over whether marriage should be promoted among soldiers as a way of cutting down on the costly incidence of venereal disease, and, if so, whether that encouragement should extend to marriages between their own male soldiers and women from the foreign countries in which they are based. Finally, regardless of which policy options are chosen, there remains the thorny question of how the military and civilian officials should describe and justify these policies to the public.[18]

Since the V–2 rockets were invented and launched over the English Channel, militaries in the most advanced industrialized societies have needed more than footsoldiers and generals. Their governments have believed they have needed underground missile silo technicians, cartographers, factory managers, engineers, and physicists.

Despite the increasing numbers of women being assigned to "non-combat" posts, most of these jobs in the militarized infrastructure are filled by men. Yet are the standards for masculinity the same in the various sectors of technology? While the frontline infantryman—or even an action-seeking lieutenant colonel sitting in the White House basement—may respond to the peculiarly Ramboesque mix of masculinity and militarization, will the scientist sitting at his laboratory computer designing a satellite laser weapon respond in the same way?

Most of our descriptions of civilian workers, business managers, scientists, and intellectuals in military contract work describe American men. For instance, William Broad has reported on the emerging subculture at the Lawrence Livermore Laboratories in

California, the site of President Reagan's Strategic Defense Initiative's ("Star Wars") most esoteric research. His book, *Star Warriors,* is not feminist—he is only slightly curious about the women whose less dramatic labors or emotional validation sustain SDI research and political lobbying. Yet he does provide considerable information about how these male scientists talk, dream, and joke.[19] Few of them are married or have girlfriends. Most of them seem to live at their computer terminals, except when they take breaks to drink coke or consume great quantities of ice cream. They don't appear to be particularly violent; they don't wear army surplus fatigues; they don't have rifles mounted on the backs of their pick-up trucks. They do have a penchant for boyish pranks. They do seem to thrive on competition and to see both the scientific world and the larger world as places where rivalry is the norm.

The latter is the ideological trait that seems to make them likely candidates for militarization. And in fact the men who recruited these young scientists made a point of playing upon their competitiveness to attract them to SDI weapons research. Still, they appear unlikely candidates for a military's basic training. They are too contemptuous of collective discipline. Their notions of action seem more cerebral than physical. And yet they clearly find deep reassurances about their own manhood in the militarized science they do. And the U.S. elites who see militarization as the bedrock of American security need for these men to feel those reassurances. That, as much as grant money, is what makes these male scientists militarizable. Admirers of Sylvester Stallone might hold these Livermore men in contempt, branding them "nerds," but they are no less critical for the 1980s militarizing of American society than M–16 wielding "grunts."

The men who lure scientists into militarized laboratory communities appear to have notions about their recruits' sexual needs not shared by their army or navy counterparts. For instance, institutionalized homophobia doesn't seem to send off warning sirens when a young male engineer eschews girls. Nor do the men responsible for the morale of the Livermore scientists seem to think they require brothels just outside the laboratory's gates. Has anyone suggested that these "Star Warriors" be sent to VD classes? Why not?

Richard Rhodes's new history of the making of the U.S. atomic bomb has been glowingly described as exhaustive and comprehensive. But when Rhodes turns his thoughts to the gendered dynamics of this militarized scientific enterprise and the peculiar community it fostered in the New Mexican desert, his curiosity suddenly wanes. Not just are the roles that women played outside his realm of historical curiosity, so are the principal male actors' notions about themselves as men interacting with women and with other men. For instance, Rhodes slips into a "light-hearted" section on the ways members of the Los Alamos community entertained themselves with a tantilizingly brief account of a debate over prostitution and "loose women." The discovery of regular visits being made by men to the single women's barracks set off an official discussion about the dangers of VD, the needs of men, and the authority of the army and civilian officials. "We did decide to continue it," one official told Rhodes. But we learn nothing more. Rhodes and his interviewee seem to share some unspoken cross-generational masculinized understanding that even in the Manhattan Project, "men will be men."[20] That understanding shuts the door on what might have been a revealing exploration into the ways in which militarized male scientists match or diverge from their infantry counterparts. Rhodes's failure of curiosity also leaves it to a future historian to uncover how the male authorities conceptualized the sexual needs of male scientists. It may be that militarized masculinity takes one form when men are socialized into the world of nuclear warfare planning, while it takes quite a different form when men are socialized into the world of what is euphemistically called conventional warfare.

Carol Cohn is a feminist who has had the unusual opportunity to spend time inside a civilian think tank at the Massachusetts Institute of Technology devoted to strategic theorizing. She describes still another breed of militarized American masculinity that has emerged since World War II—the "defense intellectual."

These men seem more likely to have social relations with women. More are married. Many of them seem able to turn on "the charm." Although all of their professional peers are men, they interact more regularly with the women who work for them as secretaries. As yet, though, we don't know much about what sorts of femininity their wives, lovers, and secretaries are pressed to adopt in order maintain smooth relationships with men who get satisfaction out of being defense intellectuals.

What Carol Cohn reveals, however, is the elaborate and often surprising linguistic and ideological formulas that these men construct to enable them to discuss death and destruction day after day. Some of the language is embarassingly (at least to the feminist ear) phallic. They talk constantly about "penetrations," "thrusts." But Cohn finds that many of the terms these American men have created to discuss efficient death and destruction are unexpectedly domestic: they pat a missile not as if it were a sexual object, but as if it were a baby or a puppy. As Carol Cohn observes, "The creatures one pats are small, cute, harmless—not terrifyingly destructive. Pat it, and its lethality disappears."[21] These men's subculture isn't quite as isolated as that of the men at Livermore Laboratory; but they have created a language and cosmology that seems to permit them to reduce their emotional investment in the actual outcomes of their abstract "scenarios."

This sort of militarized man, the defense intellectual, has come to play an increasingly influential role in those political systems that have—or wish to have—a regional or even global military role and that have technologically and industrially complex infrastructures to back up that role. It may be historically a new construction of masculinity, one that did not exist on the brink of either World War I or World War II. In societies with this brand of militarization, military planning has become too complex, presumably, for generals, politicians, and weapons producers.

Still to be written are feminist descriptions of how masculinity—and thus femininity—are constructed within companies and factories reliant on defense contracts. From Hilary Wainwright we do have a sense that at least some British women working today in the usually feminized electrical units of defense factories do not seem to get the ideological or monetary rewards that their male counterparts do from working on weapons. However, they may be encouraged to see themselves as supporting "our boys" as sort of surrogate militarized mothers.[22] But what of white-collar middle managers and senior executives, the largely masculinized management strata of aerospace, armaments, and electronics companies who have made conscious decisions to pursue local and foreign defense contracts? Perhaps they get some boost to their sense of manhood from being involved with such products and from being in regular contact with military officers.

Women as wives and lovers have to make adjustments in order to support or validate these choices. Marriages are affected when men become more and more involved in secret work, when their business days are shaped by state security concepts, or when their professional colleagues are military professionals. When a corporate contractor's wife refuses to make these adjustments in her marriage she may send small tremors through her country's military-industrial complex.

I think it would be a mistake to imagine that, because each one of these cogs in the increasingly complex military machine appears to rely on notions of masculinity, all the parts automatically work together in smooth precision. "Grunts," "nerds," "defense intellectuals," "captains of industry," "the brass"—all may be masculinized, but those masculinizations may not create bonds of trust or respect. William Broad's SDI scientists and Carol Cohn's defense intellectuals both talk of the military officers and the civilian politicians they deal with quite contemptuously. Conversely, uniformed military men of various ranks may dismiss scientists and intellectuals as too removed from reality (as they see it). It would not be surprising if some of this mutual derision were translated into comments about the other's tenuous manliness and the other's failed relations with women.

So, once again, sorting out possible varieties of masculinity—here within a single militarized society—need not induce factor-juggling paralysis. It could equip feminists with more accurate portraits of the ideological requirements any technologically sophisticated and spatially interventionist militarizing government has to try to meet. It may shed light on tensions and contradictions within those military systems, exposing them as less impermeable, more fragile.

Yet we should avoid Americanizing our analysis. It doesn't follow that every militarizing society is constructing the same divisions of labor or the same varieties of masculinity to sustain its military system. The Soviet Union, France, Britain, Israel, and South Africa may come closest to replicating the current American masculinized, militarized division of labor. Each of these states has been trying to build its own scientific and industrial military infrastructure. Each prides itself on its capacity to absorb information and make plans. But are Soviet engineers afraid of other men calling them the Russian equivalent of "nerds" if they carry calculators clipped to their belts?

Are Israeli defense intellectuals as prone as their American counterparts to looking down upon uniformed military men? They—as well as Israeli men working as engineers in the country's large weapons industry—are said to be able to sympathize with their male counterparts in uniform because they too serve in the military as reservists until the age of fifty-five. Not many societies, however, have a conscription system that helps to ensure this empathy between men socialized into different forms of militarized masculinity.

Each country may be distinguished by its own sort of tension between militarized masculinities. Consequently, roles that different groups of women are pressured to play in order to reduce or mask those tensions in the name of a smoothly running military machine might be somewhat different.

Militarization is a tricky process. It occurs both during those periods of intense militarization that we call "war" and during periods that we refer to as "peace" or "prewar" or "postwar" or "interwar." Militarization is occurring when any part of a society becomes controlled by or dependent on the military or on military values. Virtually anything can be militarized: toys, marriage, scientific research, university curriculums, motherhood, fatherhood, AIDS, immigration, racism, shopping, or comic strips. Each one of these processes involves the transformation of meanings and relationships. Rarely does it happen without the use of public power and authority. Occasionally the process is reversed. Children's play seemed to be demilitarized in the early 1970s, as evidenced by "G.I. Joe's" sharp fall in sales. Those American, Canadian, European, and Japanese scientists who are refusing to take part in SDI research are trying to demilitarize their professions. Women in any country who refuse to see their sons' accepting conscription or voluntarily enlisting in the military as a way to cope with civilian unemployment are taking steps to demilitarize motherhood. University teachers who encourage their students to assign as much analytical "seriousness" to pacifist movements as to national security policymakers are demilitarizing at least a small part of their curriculum.

Whether one is tracing militarizing social processes or demilitarizing social processes, it is necessary to chart how women and men in any particular historical setting comprehend what it means to be "manly" and what it means to be "feminine." Government and military officials not only have been effected by their own perceptions of manliness and femininity; many of them have attempted to design policies to ensure that civilians and soldiers relate to one another in those gendered ways that ease the complicated process of militarization.

Chapter 7
East-West versus North-South: Dominant and Subordinate Themes in U.S. Military Strategy since 1945

MICHAEL T. KLARE

IN THE UNITED STATES, postwar militarization occurred largely within the context of the Cold War, a complex sociopolitical phenomenon that enveloped and affected all components of American civil society. While a full assessment of the origins, nature, and consequences of the Cold War is beyond the scope of this essay, it is essential to note that one of the critical aspects of this phenomenon was a persistent national preoccupation with military preparedness—that is, with ensuring the readiness of U.S. forces and institutions to resist any type of Soviet probe or attack. Having learned from Pearl Harbor that a potential adversary might chose to initiate a powerful preemptive assault without warning or prior engagement, and believing that the Soviet Union would surely take advantage of any relaxation in U.S. defenses, American leaders sought to maintain U.S. forces and military industries at near-wartime levels of mobilization. Thus, in the name of preparedness, American society was subjected to an unprecedented degree of peacetime militarization.

This process of militarization took many forms: the creation of a large national security bureaucracy lodged in the Department of Defense, the Central Intelligence Agency (CIA), and the National Security Council (NSC); the perpetuation of military conscription and the maintenance of a large standing army, navy, and air force; the establishment of Pentagon-funded laboratories, research organizations, and "think tanks"; and the adoption of national policies entailing a high degree of public mobilization for war-related activities (like civil defense, military research and development, military training and education, intelligence and counterintelligence). Each of these developments constituted an important chapter in the history of postwar American militarization, and all of them could fruitfully be examined in a comprehensive study of this phenomenon.[1] Because of space constraints, however, I will focus here on one aspect of postwar militarization: the adoption of military strategies entailing a continuing arms competition with the Soviet Union and periodic intervention in regional Third World conflicts.

Military strategy is a vital topic for study because it represents what might be called the "commanding heights" of the militarization process. Prior to World War II, strategy formation was the exclusive pursuit of a small cadre of military professionals, and it had little, if any, peacetime impact; after the war, however, it became the paramount concern of a vastly expanded officer class and of a new corps of civilian strategists.[2] Itself the product of political and intellectual currents within this powerful constituency, strategy also had an enormous impact on civil society. Because this was a period in which military preparedness was viewed as the supreme national concern, and because civil institutions were expected to contribute as needed to the mobilization effort, changes in strategy often produced significant effects throughout the social and political system.

The powerful impact of strategic developments was due in no small measure to the introduction of nuclear weapons. Spared any of the direct military consequences of the Second World War, America suddenly found itself vulnerable to swift and unimaginable devastation—a condition that engendered intense public interest in and anxiety over things nuclear. Awed by the immense destructive power of atomic weapons and the scientific wizardry entailed in producing them, most citizens and civilian policymakers were predisposed to leave matters of nuclear policy to the strategy professionals; and when, in the name of national strategy, the professionals called for this or that military undertaking—whether it be persistent aboveground nuclear testing, the construction of new military reactors and weapons-assembly plants, or the recruitment of the nation's top scientists for classified military projects—few were inclined to offer objections.[3]

Nuclear policy was not, however, the only aspect of strategy that had a powerful impact on American society and politics. As we shall see, the question of military intervention in regional Third

World conflicts also had profound effects on civil society. Typically, the issue of intervention was addressed in the context of the U.S.-Soviet competition for world power, seen as a necessary U.S. response to Soviet-inspired insurgency on the margins of the "free world"; in reality, U.S. involvement was more often prompted by political assertiveness and indigenous militarization in areas long considered subordinate to American or Western European authority. One might say that American militarization since 1945 has contained two contradictory elements. One reflects U.S. preparedness for East-West confrontation; the other involves a readiness for intervention in the Third World that predates and may outlive the Cold War. The two interact and cannot be understood apart from one another, but it is the tension between the grand strategy of superpower politics and the realities of intervention in Third World areas that gives postwar militarization its peculiar character.

Since 1945, U.S. strategists and policymakers have grappled repeatedly with two major security issues: the nuclear arms competition with the Soviet Union and military intervention in regional Third World conflicts. These two issues have dominated the professional and academic discussion of strategic affairs, and they have constituted the basic frame of reference for the shaping, equipping, and financing of U.S. military forces in the postwar era.[4] The articulation of these issues in strategic terms has also played a significant role in domestic politics, society, and culture.

The first of these two issues I call the "major question" or the "major case" of postwar U.S. strategy. In practice, this issue breaks down into a series of subordinate questions:

(1) What policies or strategies should govern the size, orientation, and deployment of U.S. nuclear forces?

(2) Precisely what sorts of nuclear weapons and delivery systems should the United States acquire to satisfy these strategies?

(3) How should these nuclear capabilities be integrated with U.S. diplomacy in peacetime (if at all), and how should they be employed in wartime (if at all)?

(4) How and to what extent should the United States extend its "nuclear umbrella" over other countries, particularly the NATO powers of Western Europe?

(5) How should the United States seek to control, restrain, or terminate the U.S.-Soviet nuclear competition?

I call this set of questions the "major case" because together they have long been regarded as the primary security challenge facing U.S. policymakers. Because the Soviet Union also possesses a large arsenal of nuclear weapons, moreover, these questions have also tended to define the political relationship between the world's two preeminent powers. Thus, when we speak of a "summit" conference, we invariably mean a meeting between the Soviet and American heads of state, and we assume that nuclear issues will head the agenda for their conversations. And, because most analysts believe that any major war in Europe will in all probability escalate to the nuclear level, these questions also figure prominently in any discussion of European security affairs.

At the dawn of the nuclear age, U.S. strategists devised two basic approaches to the challenge of atomic warfare. The first, which is generally associated with the leaders of the U.S. Air Force (many of whom participated in the strategic bombing campaigns of World War II), held that "the bomb" was a legitimate weapon of war and that it should be used quickly and even preemptively to destroy Soviet urban-industrial centers at the first sign of an impending Soviet attack on the United States or its principal allies. To conduct such an attack—particularly once the Soviets acquired a nuclear capability of their own—Air Force strategists perceived a need for an ever-expanding arsenal of increasingly capable offensive weapons. The second approach, which is usually associated with the early writings of Bernard Brodie, held that nuclear weapons were qualitatively different from all other weapons, and they could, therefore, have only one legitimate purpose: to deter an enemy nuclear attack on the United States by posing a threat of devastating retaliation. To successfully implement such a strategy—generally known as "deterrence"—Brodie and his followers argued that the United States need only maintain a relatively small arsenal of "second-strike" weapons capable of "riding out" a first-strike nuclear attack by the USSR and then inflicting immense damage on the Soviet heartland.[5]

Although Brodie's views on deterrence enjoyed widespread support in academic and civilian policymaking circles, recently declassified Pentagon documents indicate that the air force approach generally prevailed at the Department of Defense and the Atomic Energy Commission. In line with this outlook, President Truman authorized a steady increase in America's stockpile of nuclear munitions. When in 1949 the Soviet Union demonstrated a nuclear weapons capability of its own, Truman also approved a crash program to develop a fusion-type thermonuclear weapon, the hydrogen bomb or H-bomb. And when in the 1950s the Soviets produced an H-bomb and made other improvements in their nuclear capability, President Eisenhower ordered another large increase in the U.S. nuclear stockpile.[6] As a result, the American nuclear warhead inventory rose from nine in 1946 to 450 in 1950, 2,250 in 1955, and 18,500 in 1960.[7]

As time went on, U.S. strategists introduced new refinements in nuclear strategy. Thus, in the late 1950s and early 1960s, analysts at the RAND Corporation introduced the concept of a no-cities, "counterforce" strategy, allowing U.S. authorities to counter a less-than-all-out Soviet assault with a "limited" nuclear strike against remaining Soviet nuclear weapons (hence, counterforce). To conduct such strikes, U.S. weapons would have to possess great accuracy and be responsive to exacting command-and-control arrangements. In the view of Henry Kahn, such a capability would also permit U.S. decisionmakers to devise a hierarchy of escalating responses to a wide variety of possible Soviet provocations. Similar ideas, entailing "tit-for-tat" nuclear exchanges between the Soviet Union and the United States, were advanced by Secretary of Defense James Schlesinger in the mid-1970s. Each of these strategic innovations was ultimately absorbed into official U.S. nuclear policy, thereby generating requirements for the acquisition of additional nuclear warheads and for the development of highly accurate intercontinental ballistic missiles (ICBMS).[8]

Although the United States enjoyed a substantial lead over the Soviet Union in nuclear weapons technology, it was never able to convert this advantage into lasting military superiority. Each new advance in U.S. weaponry was matched, sooner or later, by comparable advances in Soviet weaponry. This pattern led some U.S. strategists to seek fresh advantages on the frontiers of technology: thus, in the 1970s, the United States greatly increased the striking power of its Minuteman ICBMS by installing multiple warheads (MIRVS) on each missile. Other strategists, however, proclaimed the futility of such efforts and argued instead for a stable nuclear competition based on the concept of mutual deterrence, or "mutual assured destruction" (MAD).[9]

Mutual deterrence prevailed as the declared policy of the United States in the late 1970s, but meanwhile the search for new military advantages continued. In 1981, the Reagan administration embraced the concept of "protracted nuclear war," entailing repeated nuclear salvos between the two superpowers until the United States had achieved clear "victory" over the USSR. As was the case with past strategic modifications, this initiative generated added requirements for nuclear warheads and delivery systems.[10] Subsequently, Mr. Reagan introduced yet another strategic innovation: a plan for space-based strategic nuclear defense systems, or "Star Wars" as it became known. Under the Reagan plan, officially termed the Strategic Defense Initiative (SDI), the United States would deploy an elaborate system of sensors, computers, and weapons designed to identify, track, and destroy Soviet ICBMS in space before they could strike American territory. Although many scientists have questioned the viability and wisdom of deploying such a system, the Defense Department has launched a multi-billion-dollar program to develop and test the components of a Star Wars capability.[11]

Accompanying the policy debate on strategic nuclear weapons

(that is, weapons intended for strikes against the homeland of the opposing superpower) has been a comparable debate on U.S. policy regarding the possible use of so-called tactical and theater nuclear weapons in Europe. Since the Eisenhower period, U.S. strategists have viewed such weapons (which include nuclear-armed missiles, bombs, artillery shells, and demolition mines) as a necessary counterweight to the large Soviet troop concentrations in Eastern Europe. By deploying arms of this type in Western Europe, Washington has sought both to deter a Soviet invasion of Western Europe and, should deterrence fail, to provide U.S. forces with the means to liquidate enemy forces before they could completely overrun NATO's defense lines. This approach—entailing the first use of nuclear weapons by the West—became the official NATO policy in 1967 and remains so today.[12] With the signing of a U.S.-Soviet accord on the elimination of intermediate-range nuclear forces (INF) in 1987, however, Washington indicated that the degree of western reliance on tactical nuclear weapons would be significantly diminished.

In recent years, these disputes over nuclear strategy have generated a parallel debate over strategies for the control of the U.S.-Soviet nuclear arms competition. Since the Kennedy period, most U.S. policymakers have perceived some benefit (the degree to which has been a matter of continuing dispute) in negotiating mutual U.S.-Soviet limits on the size and characteristics of the nuclear arsenals of the two superpowers. This perception led to a series of arms control negotiations with the Soviet Union—a phenomenon usually described as the "SALT process," for strategic arms limitations talks. These talks in turn resulted in a number of significant superpower agreements, including the Antiballistic Missile (ABM) Treaty and SALT I accord of 1972, and the unratified (but generally observed) SALT II treaty of 1979. No further progress in arms control was made in the late Carter and early Reagan period; in 1987, however, the two superpowers agreed to a ban on the deployment of INF systems in Europe. Nevertheless, debate continues on such issues as the verification of arms control measures, Soviet compliance with existing agreements, and the perceived linkage between nuclear arms control and other East-West security issues.[13]

These disputes over nuclear policy and arms control have produced prolonged and sometimes intense debate among U.S. strategists and policymakers, but they have rarely produced outright splits in the strategic community. For the most part, disputes over nuclear policy have been reconciled through the adoption of compromise positions. Thus, in the early 1960s the debate between advocates of a no-cities, counterforce approach and those favoring an all-out punitive approach was resolved through the adoption of a nuclear target list—the so-called Single Integrated Operational Plan for Fiscal 1963 (SIOP–63)—that allowed the president to select either option. A similar debate in the Carter era between advocates of a limited-strike, tit-for-tat approach and the defenders of MAD was resolved through the adoption of Presidential Directive No. 59 (PD–59), which reaffirmed the centrality of MAD but also authorized planning for some limited-war options.[14] In the 1980s, we saw a similar debate between opponents and advocates of Star Wars—and it now appears that this dispute, too, will be resolved through a compromise allowing some research and testing of SDI components.

This process of compromise had by 1988 resulted in the development of a mammoth nuclear arsenal based on a "triad" of delivery systems—land-based ICBMs, submarine-launched ballistic missiles (SLBMs), and manned bombers—each of which, acting independently of the others, is designed to survive a Soviet attack and wreak horrendous damage on the Soviet Union. All told, the U.S. nuclear force encompassed some twenty-five thousand individual nuclear warheads, with destructive yields ranging from the equivalent of approximately ten tons to 1,200,000 tons of TNT.[15] And while U.S. policymakers continue to argue over the introduction or retirement of one or another weapons system, they have proved relatively resistant to changes in lessening the general size and organization of this awesome force.

Paralleling this tendency toward consensus and compromise at the policymaking level has been a relatively high degree of public acquiescence in elite thinking on nuclear weapons policy. Although we have seen periodic flashes of public concern over nuclear issues, the American public has for the most part acceded to prevailing government policies on weapons design, development, procurement, and deployment.[16]

There have, of course, been several significant breaches in this pattern. Probably the most important of these has been the opposition to the atmospheric testing of nuclear weapons, which boiled up in the late 1950s (when ordinary citizens learned of the buildup of strontium–90 and other radioactive materials in basic foodstuffs), and this opposition led to the atmospheric test-ban treaty of 1963. Since then, both superpowers have continued to test nuclear weapons underground—a practice that sparked a new protest movement in the mid-1980s. In addition, there was the opposition to the ABM system that peaked in the late 1960s and led to the ABM Treaty of 1972. By signing this treaty, the two superpowers essentially agreed to accept mutual vulnerability as the price of mutual deterrence. Finally, there was the opposition to the inflammatory nuclear rhetoric of the Reagan administration, as represented by Mr. Reagan's comments on the viability of a "limited" nuclear war in Europe. Such statements coupled with the massive nuclear modernization program of the Reagan administration provoked mass antinuclear demonstrations (notably the 12 June 1982 rally in New York's Central Park, attended by an estimated one million people) and gave rise to the nuclear "freeze" concept (a proposal for a bilateral freeze on the development, production, and deployment of strategic nuclear weapons). In response to these developments, Mr. Reagan toned down his rhetoric on nuclear arms and adopted a more conciliatory stance on arms control negotiations with the Soviet Union.[17]

By the end of 1986, the freeze movement had largely evaporated and public anxiety over nuclear war had greatly diminished. This anxiety has not, of course, disappeared entirely—most Americans are well aware of the terrifying consequences of nuclear war, and any new outbreak of nuclear brinkmanship would undoubtedly generate considerable public concern. But so long as the president avoids provocative language and makes a visible effort to reach accord with the Soviet Union on significant arms control measures, the general public will probably refrain from challenging basic U.S. policy on nuclear weapons.

In contrast to this pattern of relative consensus among policymakers and the general public on nuclear issues, we find a much greater degree of debate and dissent on the other major policy issue of the postwar era: the degree to which the United States should employ its military strength in resolving regional conflicts in the Third World.

As was the case with nuclear strategy, this issue also breaks down into a number of subordinate questions:

(1) Where and under what circumstances should the United States employ its own military forces in Third World conflicts?

(2) What should be the degree and duration of the U.S. military commitment to such conflicts?

(3) How should the United States respond to "gray area" challenges that lack a clear-cut threat to U.S. vital interests, and to "low-intensity" threats (including terrorism) that fall below the threshold of full-scale military hostilities?

I call this set of questions the "minor case" of strategic analysis because it has generally been viewed as a subordinate problem by U.S. strategists—in relation, that is, to the "vital threat" posed by Soviet nuclear weapons and the conventional (nonnuclear) forces of the Warsaw Pact. Thus, most formal discussions of U.S. national security policy (such as the *Annual Report* of the Secretary of Defense) devote two-thirds or more of their length to nuclear and NATO issues, leaving only minimal space for examination of what are sometimes termed "non-NATO contingencies."[18]

But while these questions have generally been regarded as secondary by U.S. strategists, they have proved highly vexing to policymakers on more than one occasion and have produced both great shifts and great schisms in public opinion. Indeed, as I will attempt to demonstrate, these questions have so deeply affected public attitudes that at least six presidential elections—those of

1952, 1960, 1964, 1968, 1976, and 1980—have been decided to a significant degree on the basis of public feelings on these issues. While it is not possible in this space to describe all of the twists and turns in the policy debate on military intervention, I would like to highlight some of the major turning points. At the end of World War II, the American public was weary of war and disinclined to support the deployment or use of U.S. troops on foreign soil. Hence, there was considerable opposition to any U.S. involvement in the Greek Civil War of 1946–49, one of the first major military challenges of the postwar era. For some policymakers, however, this antiinterventionist stance was seen as a significant obstacle to implementation of a foreign policy based on U.S. military leadership in a battered and divided world. These officials—many of whom embraced the "containment" strategy advocated by George Kennan in his "long telegram" from Moscow and the "X" article in *Foreign Affairs*—believed that the United States must be prepared to use its military strength to resist further Soviet encroachments on what was termed the free world.[19]

To overcome the perceived lethargy of the American public, advocates of a more activist posture, including President Truman, sought to portray all conflicts in the world—whatever their origin—as part of a global struggle between good and evil, freedom and tyranny. On 12 March 1947 Truman articulated this theme in a major speech on U.S. military aid to Greece and Turkey. "At the present moment in world history," he told a joint session of Congress, "nearly every nation must choose between alternative ways of life," one embracing the will of the majority and the other relying on "terror and oppression." Suggesting that the spread of communism to any corner of the globe posed a significant threat to western security, the president affirmed—in what came to be known as the "Truman Doctrine"—that "it must be the policy of the United States to support free peoples who are resisting attempted subjugation by armed minorities or by outside pressures." And while initially this principle was to be applied solely to Greece and Turkey, Mr. Truman made it clear that it would be applied to other nations when and if that proved necessary.[20]

Propelled by the crusading rhetoric of the Truman Doctrine, Congress and the public approved the president's request for substantial U.S. military assistance to the anticommunist forces in Greece. In succeeding years, many strategists sought to amend the Truman Doctrine so as to allow for the direct use of American forces (as distinct from the provision of military aid) in resisting Soviet or communist gains abroad. Indeed, such a strategy was explicitly proposed in NSC–68, a secret policy paper drawn up by Paul Nitze of the National Security Council (NSC) at the request of President Truman. Nonetheless, most Americans continued to oppose the direct use of U.S. troops in Third World conflicts. Despite much sympathy for Chiang Kai-shek, therefore, no prominent American leader was willing to suggest the use of U.S. troops to oppose communist advances in China.[21]

By 1950, however, the public mood had begun to change. Angered by heavy-handed Soviet moves in Eastern Europe and alarmed by the first Soviet nuclear tests and the communist victory in China, many public figures began to call for a more vigorous struggle against international communism. Some of this fear and anger was also directed against those Americans (especially government employees) who were thought to have somehow contributed to the success of communism abroad, either through espionage activities or by their failure to adopt a more aggressive anticommunist stance. Such paranoia, spurred on by unscrupulous politicians who sought thereby to enhance their own political fortunes (one thinks particularly of Senator Joseph R. McCarthy of Wisconsin), produced a wave of anticommunist hysteria that left a deep scar on the American political and cultural psyche. Commonly known as McCarthyism, this paranoia also generated a new commitment to the use of military force in combatting communist advances in the Third World.[22]

This commitment was soon put to the test. On 25 June 1950, North Korean forces crossed into South Korea in what appeared to be naked communist aggression against a free nation. Although many critical aspects of the Korean crisis were not considered at the time (including the possibility, supported by considerable evidence, that the South Koreans had themselves provoked the conflict), the president immediately invoked the Truman Doctrine and ordered U.S. forces to resist the North Korean attack. At first, most Americans supported the president's action, especially as it appeared that General Douglas MacArthur would engineer the West's first success in "liberating" a communist-controlled country. But when Communist China entered the conflict—thereby introducing a significant risk of nuclear confrontation—senior U.S. leaders recoiled from any further escalation of the conflict and chose instead to fight a conventional campaign restricted to the Korean Peninsula. This decision, along with the introduction of large numbers of Chinese forces, produced a bloody and frustrating stalemate on the ground, and this in turn provoked considerable public discontent in the United States. Although the discontent never resulted in antiwar protest demonstrations (as was to occur fifteen years later when a similar stalemate occurred in Vietnam), it did help elect General Dwight D. Eisenhower, who vowed in the 1952 presidential campaign to negotiate a cease-fire during his first days in office.[23]

In the wake of the Korean conflict, President Eisenhower adopted a new, scaled-down defense posture that relied to a great degree on the threat of "massive retaliation" (with nuclear weapons) to deter Soviet probes in the Third World. In line with this "New Look" policy, Mr. Eisenhower ordered a substantial cut in U.S. conventional capabilities—particularly army ground forces—as well as a significant build-up of nuclear weapons. For some citizens, this emphasis on nuclear weapons became a source of anxiety and concern (leading, in time, to the first stirrings of the ban-the-bomb movement). But most Americans generally supported the president's efforts to curb military spending and to avoid direct U.S. involvement in overseas conflicts.[24]

As the 1950s drew to a close, some American strategists began to question the logic of Eisenhower's "New Look" posture. These dissidents, led by General Maxwell D. Taylor of the army, charged that massive retaliation was an inappropriate and ineffective response to the many insurgencies and low-level military challenges that faced the United States around the world. "While our massive retaliatory strategy may have prevented the Great War," Taylor wrote in *The Uncertain Trumpet* (1959), "it has not maintained the Little Peace: that is, peace from disturbances which are little only in comparison with the disaster of general war."[25] In order to provide a credible, realistic response to such "disturbances," Taylor called for a significant expansion of America's nonnuclear forces. Such a build-up, he argued, would permit the president to implement a strategy of "flexible response"—that is, the use of whatever type of forces, nuclear or nonnuclear, that would constitute the most effective response to any challenge.

Taylor's views were embraced by then-Senator John F. Kennedy, who pledged in the 1960 presidential campaign to mount a more vigorous U.S. military response to communist probes in the Third World. This activist posture—contrasting sharply with the more restrained approach of the Eisenhower administration—was cited by many analysts as the decisive factor in Kennedy's narrow victory over Richard Nixon, then Eisenhower's vice-president. Once elected, Mr. Kennedy moved quickly to implement the proposals of Maxwell Taylor and others of his outlook. Taylor was named presidential security advisor, and flexible response was made the overarching guidance for a multibillion-dollar military build-up. In initiating this build-up, Mr. Kennedy placed special emphasis on the development of forces for "counterinsurgency" —that is, for the defeat of revolutionary guerrilla upheavals (or, as they were known at the time, "wars of national liberation"). Such movements, he believed, posed an extraordinary threat to U.S. security, and he personally supervised the development of counterinsurgency doctrine.[26]

To test the Pentagon's new weapons and tactics and to demonstrate the effectiveness of counterinsurgency strategy, President Kennedy authorized a substantial increase in the U.S. military presence in South Vietnam. In justifying this move before Congress, General Taylor, then Chairman of the Joint Chiefs of Staff, affirmed in 1963 that "here we have a going laboratory, where we see subversive insurgency, the Ho Chi Minh doctrine, being applied in all its forms." To perfect U.S. defenses against such threats, he noted, "we have recognized the importance of the area as a laboratory [and] have had teams out there looking at the equipment requirements of this kind of guerrilla warfare."[27]

Although perceived initially as a "laboratory," Vietnam soon became something rather more significant: once having been designated as a proving ground for counterinsurgency, it became essential for the United States to avoid defeat—lest America's failure in Indochina encourage revolutionaries in other countries to emulate the "Ho Chi Minh doctrine." America's credibility was put on the line and it became more and more difficult to contemplate retreat—especially when U.S. counterinsurgency efforts fell apart following the overthrow of President Ngo Dinh Diem in 1963. As suggested by Taylor in 1964,

> the failure of our programs in South Vietnam would have heavy influence on the judgments of Burma, India, Indonesia, Malaysia, Japan, Taiwan, the Republic of Korea, and the Republic of the Philippines with respect to U.S. durability, resolution, and trustworthiness. Finally, this being the first real test of our determination to defeat the communist wars of national liberation formula, it is not unreasonable to conclude that there would be a corresponding unfavorable effect upon our image in Africa and Latin America.[28]

On this basis, Mr. Kennedy and his successor, Lyndon B. Johnson, deployed more and more troops in Vietnam in what became an ever-expanding (and ultimately self-defeating) test of America's military credibility.[29]

As U.S. casualties mounted without any corresponding signs of military success, American public opinion became more and more critical of U.S. involvement in the Vietnam conflict. The growing opposition to intervention was expressed in massive student uprisings, resistance to the draft, emotional clashes between prowar "hawks" and antiwar "doves," and other symptoms of deep social malaise. At least four presidents or presidential contenders were humbled by these rifts in the body politic: Barry Goldwater lost to Mr. Johnson in the 1964 election at least in part because of public anxieties over his perceived hawkishness; Lyndon Johnson eschewed a second term in 1968 because of mounting opposition to his escalatory policies; Hubert Humphrey was defeated in 1968 at least partly because of his unwillingness to repudiate the Vietnam policies of his mentor, Lyndon Johnson; and Richard Nixon was lured into the Watergate episode and thence to his downfall by his paranoic antipathy to Vietnam doves.[30]

In the wake of Vietnam, American citizens and policymakers sought to prevent any repetition of such a fiasco by imposing a number of important restrictions on U.S. military involvement in regional Third World conflicts. These restraints, inspired by what has come to be called the "Vietnam syndrome," included the abandonment of conscription, a substantial reduction in U.S. military aid to shaky Third World governments, and a legislative ban—the War Powers Act of 1972—on the extended deployment without congressional approval of U.S. troops abroad. With these restrictions in place, Washington attempted to retain some degree of control over Third World developments by enlisting other nations—particularly Iran, Brazil, and Indonesia—to serve as regional gendarmes in America's place, an approach that came to be known as the "Nixon Doctrine."[31]

Although strongly supported by most Americans, these "post-Vietnam" restraints were opposed by some U.S. policymakers who feared Soviet exploitation of what they viewed as an American abdication of global leadership. These critics—many of whom became associated with the hard-line Committee on the Present Danger—sought to overcome public resistance to U.S. military involvement abroad by continually pointing to apparent Soviet gains in the Third World. And for a time it appeared that world events were conforming to their pessimistic assessment of the East-West power equation. Four critical events—all occurring in 1979—contributed to the perceived accuracy of this assessment:

(1) The fall of the Shah of Iran, whose departure from the scene eliminated a crucial "pillar" of the Nixon Doctrine

(2) The emergence of radical governments in Nicaragua and Grenada, which seemed to herald a new wave of guerrilla upheavals in Central America and the Caribbean

(3) The Iranian hostage crisis, which produced an emotional public outcry in the United States and generated strong demands for U.S. military retaliation against Third World terrorists

(4) The Soviet invasion of Afghanistan, which provoked a sharp anti-Soviet reaction in Washington, reminiscent in some respects to the acute hostility of the early Cold War era

These events, although largely driven by local conditions that were probably beyond Washington's ability to affect, appeared to suggest that America had somehow become impotent in the face of overseas challenges. And when President Carter failed to take the sort of aggressive action considered necessary by many Americans to cope with these challenges, he became the victim of public frustration in the 1980 election.[32]

Ronald Reagan, who had attacked Mr. Carter for his "weakness" and "vacillation" in the face of overseas pressures, vowed to restore American power and prestige when he assumed the presidency in 1981. With considerable support from Congress, he launched a five-year, $1.3 trillion program to expand and modernize America's military capabilities. A significant fraction of this largesse—approximately 25 percent—was devoted to nuclear weapons; the bulk of it, however, was devoted to conventional weapons, including aircraft carriers and other "power projection" forces intended primarily for use in Third World areas. Reagan also called for the revitalization of America's special operations forces (the army's Special Forces, the navy's SEALs, and other such detachments), and he approved other measures intended to enhance the U.S. capability for military intervention in regional Third World conflicts. Citing strong public support for these initiatives, Mr. Reagen averred in 1981 that the Vietnam syndrome had been cured and that "the people of America have recovered from what can only be called a temporary aberration."[33]

Accompanying this build-up in interventionary capabilities, the

Reagan administration also approved a thorough overhaul of U.S. strategy for counterinsurgency and other forms of "low-intensity conflict" (LIC). Tactics developed during the Vietnam era were modified to fit contemporary realities and then extensively field-tested in El Salvador. As the 1980s proceeded, moreover, the Pentagon's LIC doctrine was expanded to include such activities as counterterrorism, the use of military forces to combat drug smuggling, and the initiation and support of antigovernment insurgencies in Nicaragua, Angola, and other Third World countries ruled by pro-Soviet regimes.[34] This latter effort, popularly known as the "Reagan Doctrine," became one of the most conspicuous and controversial policy initiatives of the president's second term, when administration efforts to support the Nicaraguan *Contras* with proceeds from secret arms sales to Iran resulted in a major national scandal.[35] Despite the dramatic impact of the Iran-Contra scandal, however, U.S. policymakers have generally retained their commitment to the reinvigoration of America's LIC capabilities.

This brings us more or less up to the present. Having completed this journey through the landscape of postwar U.S. strategic thinking, one is struck by the degree to which nonnuclear issues—particularly issues raised by U.S. military involvement in Third World conflicts—have dominated the foreign policy debate and shaped the public mood. Indeed, the very prominence of these North-South conflict issues appears to contradict the common wisdom that East-West security issues constitute the "major case" of U.S. security policy. This, in turn, leads to two important questions: (1) is this assessment of the relative preeminence of the "minor case" accurate?; and (2) if so, why is this the case?

In attempting to answer these questions, it is necessary to begin with the observation that it is very difficult to gauge the impact of nuclear weapons issues on American society because, as noted by many psychologists, the fear of nuclear war can be so overpowering to the human psyche that people often suppress their feelings about the bomb and tend to behave as if the problem is very remote from their everyday concerns. One also has to acknowledge the "cult of secrecy" that surrounds nuclear weapons issues, as well as the great complexity of the issues involved (a condition that tends to intimidate ordinary citizens and keep them from participating in the public debate) and the fact that there have not been any instances of nuclear combat since 1945.[36] In contrast, nonnuclear conflicts (of which there have been many since 1945) are relatively susceptible to public evaluation and they tend to produce strong emotions—whether of rage, patriotism, or revulsion—that are easier for the average citizen to express and cope with.

All this having been said, one is still confronted with the fact that North-South conflict issues appear to have had a much more powerful impact on the political process over the past forty years than has the East-West conflict. One need only recall the public outrage unleashed by the seizure of U.S. hostages in Iran and the emotional response to the hostages' return to appreciate the manner in which such North-South issues have dominated the political landscape. Indeed, while I can think of no presidential election that has been determined to any significant degree by East-West issues, I believe that six such elections have been decided to an important degree by North-South issues.

What, then, accounts for the apparent predominance of the "minor case"? While it is impossible to provide a thorough and complete answer to this question here, several factors appear to be critical: To begin with, it is necessary to acknowledge that the Third World has in fact become the only safe battleground for the active expression of East-West antagonisms. Because any military confrontation between the United States and the Soviet Union is likely to result sooner or later in nuclear retaliation, the superpowers have tended to sublimate their hostility to one another by taking opposing sides in regional Third World conflicts involving their allies and clients. Thus it often occurs that local Third World conflicts with largely indigenous origins—one thinks, for instance, of the Korean War—become much more violent and conspicuous than they might otherwise have been because of the direct or indirect involvement of the superpowers. As suggested by Marshall D. Shulman in 1986,

> the importance of [Third World conflict issues] stems from the fact that, although the regulation of nuclear weapons may have the most immediacy and urgency on the United States–Soviet agenda, the conflicts that arise in the Third World . . . may be more likely to engage the superpowers in a dangerous confrontation than the intersection of their interests in Europe, which they have learned over the decades to handle with great caution.[37]

In this sense, many North-South conflicts can be seen as surrogates for rather than departures from East-West conflicts.

On the other hand, it is important to note that the United States has often intervened in Third World conflicts for reasons that appear to bear little relevance to the East-West rivalry. Thus, the United States exercised *de facto* dominion over Latin America long before the Bolsheviks rose to power in Russia, and what we are witnessing today in Central America appears to have more in common with U.S. "peacekeeping" in that region during the early 1900s than with the sort of "surrogate" conflict seen in Korea.[38] Furthermore, the United States has in the past forty years acquired major economic interests in the Third World, ranging from the oilfields of the Middle East to the mineral supplies of Africa and the industrial zones of the Pacific rim. In these areas, it is not so much Soviet-inspired communism that threatens U.S. interests as it is home-grown nationalism, radicalism, and religious zealotry. Thus, while Washington may claim that American military involvement in such areas is largely inspired by a desire to resist Soviet aggression, the dominant motive is more likely a fear of indigenous revolutionary movements.[39]

It would also appear that time has taken some of the sting out of the U.S.–Soviet antagonism. While it is certainly true that anti-Sovietism was a potent political force in the McCarthy period, and while most Americans continue to fear and dislike the Soviet system, there seems to be a long-term decline in the degree of risk-taking that the public is willing to accept in the military dimension of the U.S.–Soviet competition. Thus, while Congress and the public supported President Reagan's drive to stiffen U.S. defenses against Soviet military power, they were much less enthusiastic about his inflammatory anti-Soviet rhetoric and his rather cavalier attitude toward arms control. Indeed, many analysts believe that Mr. Reagan was able to ensure a clear-cut victory in the 1984 election only by softening his rhetoric and by promising to pursue new arms control initiatives with the Soviet Union.[40] Subsequently, when the president met with Mikhail Gorbachev of the Soviet Union and pledged to work for the elimination of nuclear weapons, the renewed Cold War atmosphere that pervaded Washington in the early 1980s began to dissipate. And while it is too early to assess the full impact of Mr. Gorbachev's reform efforts, his policies of *glasnost* (or "openness") appear to have engendered a more sympathetic U.S. public attitude toward the Soviet Union.

This relaxation in U.S. concern over the East-West conflict has not, however, been accompanied by a similar shift regarding the North-South conflict. Indeed, while public sympathy for Mr. Gorbachev appears quite substantial, most Americans harbor a continuing hostility toward Colonel Qaddafi of Libya and the Ayatollah Khomeini of Iran—maverick rulers who are seen as the masterminds of a virulent anti-American crusade that is sweeping through the Third World. Possibly because terrorism is experienced as a personal (as distinct from societal) threat by American citizens, or because Third World insurgencies can stir up latent feelings of ethnocentricism and racism, North-South conflict issues have tended to provoke a sharper emotional response in recent years than have comparable East-West issues. While most Americans continue to identify the Soviet Union as posing the greatest potential threat to U.S. security, they are far more likely to advocate military action in responding to threats arising in the Third World.[41]

While this greater readiness to contemplate the use of force in combating Third World challenges obviously reflects a keen appreciation of the escalatory risks inherent in any confrontation with the Soviet Union, it also stems from a largely unspoken belief that in the final analysis America's conflicts with hostile Third World nations and forces may prove more intractable than the conflict with the Soviet Union. While U.S.–Soviet hostility embodies powerful ideological differences, there are few concrete interests now separating the two superpowers: the Soviet domination of Eastern Europe—once the cause of great friction between East and West—has largely been accepted through the Helsinki Accords of 1975, and the Soviet occupation of Afghanistan was terminated by Mikhail Gorbachev in 1988–89. On the other hand, the United States faces continuing and possibly violent disputes with Third World countries over vital geoeconomic concerns, including basing rights, maritime transit rights, overseas investments, and access to critical sources of strategic raw materials. Such issues have arisen in almost every previous instance of U.S. intervention in the Third World, and such conflicts are likely to erupt with increasing intensity in future years as these countries seek greater control over their economic development.

This perception is rarely articulated openly in the strategic literature. In 1977, however, these concerns were raised in an influential study by the RAND Corporation entitled *Military Implications of a Possible World Order Crisis in the 1980s*. There is a growing risk, the report notes, that humanity "faces the possibility of a breakdown of the global order as a result of a sharpening confrontation between the Third World and the industrial democracies." Because of a growing gap between rich and poor, "the North-South conflict . . . could get out of hand in ways comparable to the peasant rebellions that in past centuries engulfed large parts of Europe or Asia, spreading like uncontrollable prairie fires." And because only the United States has the military might to contain such conflagrations, it will "be expected to use its military force to prevent the total collapse of the world order or, at least, to protect specific interests of American citizens."[42]

Later in the 1980s similar statements appeared in the military literature on low-intensity conflict, terrorism, and counterinsurgency. In a 1983 speech at the National Defense University, Pentagon adviser Neil C. Livingstone argued that "unfulfilled expectations and economic mismanagement have turned much of the developing world into a 'hothouse of conflict,' capable of spilling over and engulfing the industrial West." Because this disorder threatens "the survival of our country and way of life," U.S. security "requires a restructuring of our warmaking capability, placing new emphasis on the ability to fight a succession of limited wars and to project power into the Third World."[43] These views were later echoed by Lt. Colonel Oliver North, a protégé of Livingstone, during the congressional hearings on the Iran-Contra affair.

This perspective—that the United States is vitally threatened by insurgent forces in the Third World and that military force is needed to resist this threat—gained increased currency in the mid-1980s. Recognizing that the likelihood of armed conflict with the Soviet Union had greatly diminished, many U.S. strategists began to argue that America faced a more critical peril in the developing areas—a peril that had been inadequately addressed in previous years because of an overarching concern with the Soviet–Warsaw Pact threat. Thus, in a 1985 article in *Military Review* (the theoretical journal of the army's Command and General Staff College), Colonel James Motley affirmed that "the United States should reorient its forces and traditional policies from an almost exclusive concentration on NATO to better influence politico-military outcomes in the resource-rich and strategically located Third World areas."[44]

This outlook, which would have been considered heretical by most strategists in the early 1980s, became established doctrine in 1988 with the publication of *Discriminate Deterrence,* the report of the Commission on Integrated Long-Term Strategy (an advisory body created by the Department of Defense and the NSC in 1987). "An emphasis on massive Soviet attacks leads to tunnel vision among defense planners," the report notes. "An excessive focus on these contingencies diverts defense planners from trying to deal with many important and far more plausible situations," especially conflicts arising in the developing areas. In the commission's view:

> These conflicts in the Third World are obviously less threatening than any Soviet-American war would be, yet they have had and will have an adverse cumulative effect on U.S. access to critical regions, on American credibility among allies and friends, and on American self-confidence. If this cumulative effect cannot be checked or reversed in the future, it will gradually undermine America's ability to defend its interests in the most vital regions, such as the Persian Gulf, the Mediterranean and the Western Pacific.[45]

Given this assessment, "In the coming decades the United States will need to be better prepared to deal with conflicts in the Third World."

This outlook is likely to prevail in the 1990s. It will prevail because it reflects underlying geoeconomic concerns and because U.S. strategists perceive a growing threat from hostile forces in the Third World. This perception reflects not only a sense that such threats are multiplying, but also that prospective Third World adversaries are much better equipped and organized than those in the past, because of the massive shift in military capabilities (arms, training, and technical know-how) from North to South in recent decades.[46]

As the perceived threat from the Soviet Union decreases and as Third World countries perceived as being hostile acquire sophisticated arms, it is natural to expect a strategic revision of the sort now underway in the United States. But something else appears to be at work here: as John Gillis notes in his introduction to this volume, the process of militarization consistently requires a threatening "other" to justify the unceasing commitment of national resources to military purposes. In the immediate postwar era, the role of the "other" was assigned to the Soviet Union and, after 1949, to Communist China; today, with the ascendancy of reform-minded regimes in both Moscow and Beijing, it is much harder to make this assignment, and thus there arises the tendency to elevate the threat posed by menacing Third World figures like Colonel Qaddafi and the Ayatollah Khomeini.

Indeed, these figures serve as stand-ins for an even more threatening "other"—the nameless masses of the Third World who supposedly covet America's wealth and reject America's value systems. In perhaps the most explicit expression of this characterization, General Maxwell Taylor wrote in 1974 that, "As the leading affluent 'have' power, we may expect to have to fight to protect our national valuables against envious 'have-nots.'"[47] While expressed in terms that every American could understand, this statement draws on the same concern with the maintenance of privilege that fueled the militarization of Europe in the first decades of this century.[48] This concern appears to lie deep in the American psyche, and it is perhaps for this reason that North-South issues have erupted so often and with such unpredictable results in last forty years.

Notes

Introduction

1. The 1986 edition of the *International Encyclopedia of the Social Sciences* defines militarism as "a doctrine or system that values war and accords primacy in state and society to the armed forces" (vol. 10, p. 300). It does not acknowledge the concept of militarization. For the development of concept of militarism, see Volker R. Berghan, *Militarism: The History of an International Debate, 1861–1979* (Cambridge: Cambridge University Press, 1984).

2. This is Michael Geyer's definition. See p. 79 below.

3. The *Oxford English Dictionary* dates "militarism" to the 1860s, with "militarization" being used first in the 1880s. The latter term, however, was not applied to societies at large until it was used by contemporary social scientists.

4. Michael Mann, "The Roots and Contradictions of Modern Militarism," *New Left Review* 162 (March/April 1987):40.

5. Berghahn, *Militarism,* chapter 4.

6. For suggestions on how this gap might be bridged, see Stephen Wilson, "For a Socio-Historical Approach to the Study of Western Military Culture," *Armed Forces and Society* 6, no. 4 (Summer 1980): 527–552.

7. A useful example is J. A. Mangan and James Walvin, eds., *Manliness and Morality: Middle-Class Masculinity in Britain and America, 1880–1940* (Manchester: University of Manchester Press, 1987).

8. An outstanding exception is Claudia Koonz, *Mothers in the Fatherland: Women, the Family and Nazi Politics* (New York: St. Martin's, 1986); also Margaret Higonnet, Jane Jenson, Sonya Michel, Margaret Weitz, eds., *Behind the Lines: Gender and the Two World Wars* (New Haven: Yale University Press, 1987).

9. Robin Luckham, "Armament Culture," *Alternatives: a Journal of World Policy* 10, no. 1 (Summer 1984):1.

10. On the international dimensions of militarization, see Asbjorn Eide and Marek Thee, eds., *Problems of Contemporary Militarism* (London: Croom Helm, 1980).

11. Mann, "Roots," pp. 47–49.

12. Eric Leed, *No Man's Land: Combat and Identity in World War I* (New York: Cambridge University Press, 1979); John Keegan, *Face of Battle* (New York: Vintage Books, 1987).

13. This is amply demonstrated by Eugen Weber's *Peasants into Frenchmen: The Modernization of Rural France, 1870–1914* (Stanford: Stanford University Press, 1976); for a useful criticism of the literature that takes the nation-state for granted, see Michael Mann, *The Sources of Social Power,* 2 vols. (Cambridge: Cambridge University Press, 1986), 1:1–33; For a useful global perspective, see Michael Geyer and Charles Bright, "For a Unified History of the World in the Twentieth Century," *Radical History Review* 39 (September 1987):69–91.

14. Mann, "The Roots and Contradictions of Modern Militarism," pp. 45–47.

15. Figures from Marek Thee, "Militarization in the United States and the Soviet Union: The Deepening Trend," *Alternatives: A Journal of World Policy* 10, no. 1 (Summer 1984):103.

16. Wilson, "Socio-historical Approach," pp. 541–544.

17. Thee, "Militarization," p. 110.

Chapter 1.
The Militarization of European Society, 1870–1914

1. Goldsworthy Lowes Dickinson, *The International Anarchy* (London: Allen and Unwin, 1926); Henry N. Brailsford, *The War of Steel and Gold. A Study of the Armed Peace* (London: Bell, 1914). This "indian summer" atmosphere has been admirably expressed by Barbara Tuchman in *The Proud Tower: a Portrait of the World Before the War* (London: Hamilton, 1966).

2. *War by Timetable* was taken as a title for one of A. J. P. Taylor's several good short books about on World War I. (London: Macdonald, 1969). The total impact of the war upon the belligerents is well summarized by Marc Ferro, *The Great War 1914–1918* (London: Routledge and Kegan Paul, 1973) and Keith Robbins, *The First World War* (Oxford: Oxford University Press, 1985). An unusually comprehensive survey of what the war meant to one county, Great Britain, is given by Trevor Wilson, *The Myriad Faces of War* (Cambridge: The Polity Press, 1986). Sixty million is only an approximation but that is what the na-

tional figures add up to. Russian lost an estimated twelve million, Germany eleven million, France eight million, Austro-Hungary seven million plus, Britain and its empire nine million, Italy five million plus, the United States 4 million plus, Turkey three million, Bulgaria one million plus.

3. For these German aspects, see the essays of Wolfgang Petter and Wilhelm Diest in *The German Military in the Age of Total War,* ed. Wilhelm Diest (Leamington, England: Berg, 1985).

4. A class survey of conscription is V. G. Kiernan, "Conscription and Society in Europe before the War of 1914–18," in *War and Society: Historical Essays in Honour and Memory of J. R. Western,* ed. M. R. D. Foot (London: Paul Elek, 1973), pp. 141–158. For French facts and conclusions, I am indebted to Douglas Porch, "L'armée française et l'esprit offensive 1900–1914," a paper delivered at a Colloque at the Institut d'Etudes Politiques de Toulouse, September 1976. For the Russian references, see Norman Stone, *Europe Transformed, 1878–1919* (London: Fontana, 1983), pp. 223, 353; for German references, see the same, p. 332.

5. Porch, "L'armée française." Alan Sked's figures for the years 1876 to 1890 show a rate of 12.53 per ten thousand men in the Hapsburg army; just over 8 in the next most suicidal army, that of Saxony and Wurtemberg; 3.33 in France, and 2.09 in the United Kingdom. Further glimpses of the recruiting deficiences of the Hapsburg army appear in Norman Stone, "Army and Society in the Hapsburg Monarchy, 1900–1914," *Past and Present* 33 (1966):95–111.

6. Marder's estimate is cited in Robbins, *The First World War,* p. 85; also Maurice Pearton, *The Knowledgeable State: Diplomacy, War and Technology since 1830* (London: Burnett, 1982); William McNeill, *The Pursuit of Power: Technology, Armed Force and Society since A.D. 1000* (Oxford: Blackwell, 1985).

7. Karl Demeter, *The German Officer Corps in Society and State, 1640–1945* (London: Weidenfeld, 1965); information from Alan Sked; John Gooch, *Armies in Europe* (London: Routledge and Kegan Paul, 1980), p. 139; Gunther Rothenberg, *The Army of Franz Joseph* (West Lafayette: Purdue University Press, 1976).

8. Cynthia Enloe, *Does Khaki Become You? The Militarization of Women's Lives* (London: Pluto, 1983); David Mitchell, *Women on the Warpath* (London: J. Cape, 1966); Arthur Marwick, *Women at War, 1914–1918* (London: Fontana, 1977).

9. Anna Davin, "Imperialism and Motherhood," *History Workshop*

Journal 5 (1978):9–65; useful on militarism generally, but less informative about women in particular is Ann Summers, "Militarism in Britain before the Great War," *History Workshop Journal* 2 (1976): 104–123. Also Margarethe Ludendorff, *My Married Life* (London: Hutchinson, 1930); Robert Graves, *Goodbye to All That* (London: Penguin Books, 1960), pp. 188–191.

10. The Moltke passage, often cited, may be found for instance in Geoffrey Best, *Humanity in Warfare* (London: Weidenfeld, 1980), pp. 144–45.

11. I. F. Clarke, *Voices Prophesying War, 1763–1984* (London: Oxford University Press, 1966); "Such Good Copy," in Max Beerbohm, *Fifty Caricatures* (London: Heinemann, 1913), no. 34.

12. For British youth, see especially John Springhall, *Youth, Empire and Society: British Youth Movements, 1883–1940* (London: Croom Helm, 1977); Trevor Wilson, *Myriad Faces of War*, p. 706, cites the popular recruiting song (performed by Maggie Smith in the film *Oh What a Lovely War!*) which began: "On Sunday I walk out with a soldier, On Monday I'm taken by a tar; On Tuesday I'm out with a baby Boy Scout, On Wednesday with a Hussar . . ." For the German equivalents, see Walter Laqueur, *Young Germany* (London: Routledge and Kegan Paul, 1962) and Peter D. Stachura, *The German Youth Movement 1900–1945* (London: Macmillan, 1981).

13. See especially Geoff Eley, *Reshaping the German Right* (New Haven: Yale University Press, 1980) and Paul Kennedy and A. J. Nicholls, eds., *Nationalist and Racialist Movements in Britain and Germany before 1914* (London: Macmillan, 1981).

14. For the 1899 conference whose origins made it politically and ideologically the more interesting of the two, there has never been a better study than that of Albert Geouffre de Lapradelle, *La Conference de La Paix* (Paris: Pedone, 1900).

15. Jost Dülffer, *Regeln gegen den Krieg? Die Haager Friedenskonferenzen 1899 und 1907 in der internationalen Politik* (Berlin: Ullstein, 1981). Joylyon Howorth, "French Workers and German Workers: the impossibility of internationalism, 1900–1914," *European History Quarterly* 15 (1985):71–97.

16. Robbins, *The First World War*, p. 17.

17. Jean-Jacques Becker, *The Great War and the French People* (Leamington, England: Berg, 1985). For the Russian reference, see Allen K.

Wildman, *The End of the Russian Imperial Army* (Princeton: Princeton University Press, 1980), pp. 75–80. I am also indebted to a conversation with Peter Simpkins.

Chapter 2.
Militarization and Rationalization in the United States, 1870–1914

1. Stephen Wilson, "For a Socio-Historical Approach to the Study of Western Military Culture," *Armed Forces and Society* 6 (1980): 527–552. See Volker R. Berghahn, *Militarism: The History of an International Debate, 1861–1979* (New York: St. Martin's Press, 1982), for a good account of the varied meanings of the term "militarism" over the past century and a quarter.

2. See Peter Karsten, "Armed Progressives: The Military Reorganizes for the American Century," in *Building the Organizational Society,* ed. Jerry Israel, (New York: Free Press, 1972), pp. 197–232.

3. See B. F. Cooling, *Grey Steel and Blue Water Navy: The Formative Years of America's Military-Industrial Complex* (Hamden, Conn.: Archon, 1979); B. F. Cooling, ed., *War, Business, and American Society: Historical Perspectives on the Military-Industrial Complex* (Port Washington, N.Y.: Kennikat, 1977); and Peter Karsten, *The Naval Aristocracy: The Golden Age of Annapolis and the Emergence of Modern American Navalism* (New York: Free Press, 1972), pp. 174–185.

4. Samuel P. Huntington, *The Soldier and the State* (Cambridge, Mass.: Harvard University Press, 1957); J. P. Mallan, "Roosevelt, Brooks Adams, and Lea: The Warrior Critique of the Business Civilization," *American Quarterly* 8 (1956): 216–230.

5. See Wallace E. Davies, *Patriotism on Parade* (Cambridge, Mass.: Harvard University Press, 1955) for evidence regarding the phenomenon within the Grand Army of the Republic into the 1880s and 1890s, as well as in such patriotic organizations as the Sons and Daughters of the American Revolution and the Society of American Wars. See also notes 12, 13, 14, 16, and 26.

6. Gerald Linderman, *Embattled Courage: The Experience of Combat in the American Civil War* (New York: The Free Press, 1958), p. 284; Lloyd Lewis, *Sherman: Fighting Prophet* (New York: Harcourt Brace, 1958), pp. 457, 515, 637–47.

7. Holmes, "The Soldier's Faith," in *The Occasional Speeches of Justice Oliver Wendell Holmes,* ed. Mark DeWolfe Howe (Cambridge, Mass.: Harvard University Press, 1962), pp. 755–782.

8. Davies, *Patriotism on Parade,* pp. 75, 261, 315, 333–341. Thurston address is in *Addresses delivered at the Lincoln Dinners of the Republican Club of the City of New York,* 5 vols. (New York: private printing 1897–1903), 1:95.

9. U. S. Congress, House of Representatives, *Congressional Record* 3 May 1916, speech of Representative Tavenner; *Army and Navy Journal* 47 (4 September 1909):15; 48 (12 November 1910):299; Hudson Maxim, *Defenseless America* (New York: Hearst's International Library 1915), pp. 136, 138, 140.

10. Karsten, *Naval Aristocracy,* chapters 5, 6, and 7. Irwin Wyllie found that few American businessmen made use of Darwinistic metaphors in the fin de siècle, but my evidence shows that U.S. naval officers in these same years were conversant with Darwinism and that most of them could fairly be styled "national Darwinists."

11. See Robert A. Hart, *The Great White Fleet* (Boston: Little Brown, 1965), p. 67.

12. See Karsten, *Naval Aristocracy,* pp. 209, 218, 257, 266–267.

13. See Allan Millett and Peter Maslowski, *For the Common Defense* (New York Free Press, 1984), p. 323; Theodore Ross, *Odd Fellowship: Its History and Manual* (New York: M. W. Hazen, 1887), pp. 440–460; James Malin, *Confounded Rot about Napoleon* (Lawrence, Kan.: Regents, 1961), pp. 185–204; Theodore Greene, "The Hero as Napoleon," in Theodore Greene, *America's Heros* (New York: Oxford University Press, 1970); and Peter Karsten, *Patriotic-Heroes in England and America: Political Symbolism and Changing Values over Three Centuries* (Madison, Wisconsin: University of Wisconsin Press, 1978), chapter 4, on the evidence of veneration of both Napoleon and Oliver Cromwell in America in these years.

14. William James, "The Moral Equivalent of War," in Staughton Lynd, *Non-Violence in America: A Documentary History* (Indianapolis: Bobbs-Merrill, 1966), pp. 136–160; James to Fredric Myers, 1 January 1896, and to Theodore Fournoy, 17 June 1898, in Ralph Barton Perry, *The Thought and Character of William James,* 2 vols. (Boston: Little, Brown, and Co., 1935–1936), 2:305, 308.

15. Arthur Lipow, *Authoritarian Socialism in America: Edward Bellamy and The Nationalist Movement* (Berkeley and Los Angeles: Univer-

sity of California Press, 1982), pp. 37, 39; Edward Bellamy, "What Nationalism Means," *The Contemporary Review* (July 1890), reprinted in *Edward Bellamy Speaks Again!* (Kansas City: Peerage Press, 1937), pp. 92, 94; Edward Bellamy "An Echo of Antietam," *Century Magazine* (July 1889); John L. Thomas, *Alternative America* (Cambridge, Mass: Harvard University Press, 1983), pp. 266, 273; *Edward Bellamy Speaks Again,* pp. 188–189.

16. Thomas A. Bailey, *A Diplomatic History of the American People,* 10th edition (Englewood Cliffs, N.J.: Prentice Hall, 1980), p. 461; *Washington Post,* 2 June 1898.

17. David McClelland, ("Love and Power: Psychological Signals of War," *Psychology Today* [January 1975]:44ff) and his associates content-analyzed American children's literature over the past two-and-a-half centuries and found a cyclical correlation between essays in that literature designed to inspire an appreciation for power, as in the late nineteenth century, and the outbreak of war. Similarly, Frank L. Klingberg ("The Historical Alternation of Moods in American Foreign Policy," *World Politics* 9 [1955]:239–273) found similar cyclical phases of aggressiveness and militarization in American foreign policy, correlated with such phenomena as bellicosity in presidental addresses and in the platform of the successful political party, and naval expenditures. He found the period 1892 to 1916 to be such an aggressive era.

18. Wilson, "Socio-Historical Approach," pp. 528, 546.

19. J.A.S. Greenville and G. B. Young, "The Influence of Strategy upon History: The Acquisition of the Philippines," chapter 10 of their *Politics, Strategy and American Diplomacy* (New Haven: Yale University Press, 1966); Ronald Spector, "Who Planned the Attack on Manila Bay?" *Mid-America* (April 1971):93–104.

20. From an editorial in the *American Tribute* (Indianapolis) 10 June 1897. I will say little of the possibility that Richard Hofstadter's "psychic crisis" theory provides evidence of war lust in the 1890s. See Hofstadter, "Manifest Destiny and the Philippines," in *America in Crisis,* ed. Daniel Aaron (New York: Alfred A. Knopf, 1952). Hofstadter claimed that elites like President Grover Cleveland and his Secretary of State Richard Olney sensed the potential for distracting the nation from its social and economic woes during the mid-1890s depression by provoking Great Britain at the risk of war. However, he provided very little evidence that such views were held by policymakers, and his theory has not been corroborated by new evidence. Moreover, even if such evidence were to be

unearthed, it would not establish that war lust was widespread among the populace; it would only provide new evidence that some elites believed that it was.

21. Walter LaFeber, *The New Empire: An Interpretation of American Expansion, 1860–1898* (Ithaca, N.Y.: Cornell University Press, 1963), pp. 350–410.

22. James L. Abrahamson, *America Arms for a New Century: The Making of a Great Military Power* (New York: Free Press, 1981), p. 43.

23. John Schofield, *Forty-six Years in the Army* (New York: Century Company, 1897), p. 457. See also Lester Langley, "The Democratic Tradition and Military Reform, 1878–1885," *Southwestern Social Science Quarterly* (1967):192–200.

24. See Richard Werking, *The Master Architects: Building the United States Foreign Service, 1890–1913* (Lexington, Ky.: University of Kentucky Press, 1977).

25. See Samuel P. Hays, *Conservation and the Gospel of Efficiency* (Cambridge, Mass.: Harvard University Press, 1959).

26. See Thomas McCormick, *China Market: America's Quest for Informal Empire, 1893–1901* (Chicago: Quadrangle Books, 1967).

27. See Michael Lutzker, "The Pacifist as Militarist: A Critique of the American Peace Movement, 1898–1914," *Societas* 5 (Spring 1975): 87–104.

28. Gifford Pinchot to Samuel Gompers, 10 March 1917, cited in David M. Kennedy, *Over Here: The First World War and American Society* (New York: Oxford University Press, 1980), p. 146.

29. See J. Garry Clifford, *The Citizen Soldiers: The Plattsburg Training Movement, 1913–1920* (Lexington, Ky.: University of Kentucky Press, 1972); J. P. Finnegan, *Against the Spector of a Dragon: The Campaign for Military Preparedness, 1914–1917* (Westport, Conn.: Greenwood Press, 1974); John Chambers, *To Arm a Nation: The Adoption of Conscription in America* (New York: Free Press, 1987); and Robert Cuff, *The War Industries Board: Business-Government Relations During World War I* (Baltimore: John Hopkins University Press, 1973).

30. Quoted in John P. Finnegan, "Military Preparedness in the Progressive Era, 1911–1917" (Ph.D. dissertation, University of Wisconsin, 1969), pp. 43, 49–50.

31. Figures from Harold Wool, *The Military Specialist* (Baltimore: Johns Hopkins University Press, 1968), p. 2.

32. Hudson Maxim, *Defenseless America*, p. 169; for additonal fig-

ures, see Harold and Margaret Sprout, *Rise of American Naval Power, 1776–1918* (Annapolis: Naval Institute Press, 1966), pp. 290–303; Fletcher Pratt, *The Compact History of the U.S. Navy,* rev. ed., (New York: Hawthorn Books, 1962), p. 196.

33. Walter Millis, *Arms and Men* (New York: Putnam, 1956), p. 176 n. 1.

34. Russell Weigley, *The American Way of War* (New York: Macmillan, 1973), p. 196.

35. John Chambers, "Conscription for Colossus: The Progressive Era and the Origin of the Modern Military Draft in the U.S. in World War I," in *The Military in America From Colonial Times to the Present,* 2nd ed., ed. Peter Karsten (New York: Free Press, 1986), pp. 297–311.

36. Marcus Cunliffe, *Soldiers and Civilians: The Martial Spirit in America, 1775–1865* (New York: Free Press, 1976), chap. 10.

37. Neither, of course, did Britain or Japan, but both of those nations were more militaristic than the United States and Japan was one of the most militaristic states in the fin de siècle. There the survival of feudal values and a sense of danger from western powers joined in ways quite foreign to the United States.

38. Fredric Bancroft, ed., *Speeches, Correspondence and Political Papers of Carl Schurz,* 6 vols. (New York: G. P. Putnam, 1913), 5:207, 258–259, 264–265.

39. Maxim, *Defenseless America,* pp. 133–134.

40. Ibid., p. 113.

Chapter 3.
Toward a Warfare State

1. The analysis that follows is based primarily, although not exclusively, on my published works: *The Hammer and the Sword: Labor, the Military, and Industrial Mobilization, 1920–1945* (New York: Arno Press, 1979); *The Military-Industrial Complex: A Historical Perspective* (New York: Praeger Publishers, 1980); and "Warfare and Power Relations in America: Mobilizing the World War II Economy," in *The Home Front and War In the Twentieth Century: The American Experience in Comparative Perspective,* ed. James Titus (Colorado Springs, Colo.: United States Air Force Academy, 1984), pp. 91–110, 119–120, 231–243. The last two works contain elaborate bibliographies relevant

to this subject. Additionally, I am drawing upon my work that is still in progress and will be published in two volumes: "Beating Plowshares Into Swords: The Political Economy of Warfare in America."

In the text that follows, I will only cite works that I believe are especially important and that are not included in the publications cited above.

2. The discussion of the preparedness movement is based on: John W. Chambers, *To Raise an Army: The Draft comes to Modern America* (New York: Free Press, 1987); Kendrick A. Clements, *William Jennings Bryan: Missionary Isolationist* (Knoxville, Tenn.: University of Tennessee Press, 1983); John Milton Cooper, Jr., "Progressivism and American Foreign Policy: A Reconsideration," *Mid-America* 51 (October 1969): 266–277; John Milton Cooper, *The Vanity of Power: American Isolationism and the First World War, 1914–1917* (Westport, Conn.: Greenwood Press, 1969); Allen F. Davis, "Welfare, Reform, and World War I," *American Quarterly* 19 (Fall 1967):516–533; Arthur A. Ekirch, Jr., *Progressivism in America: A Study of the Era from Theodore Roosevelt to Woodrow Wilson* (New York: New Viewpoints, 1974); John Patrick Finnegan, *Against the Specter of a Dragon: The Campaign for American Military Preparedness, 1914–1917* (Westport, Conn.: Greenwood Press, 1974); Otis L. Graham, Jr., *An Encore for Reform: The Old Progressives and the New Deal* (New York: Oxford University Press, 1967); Charles Hirschfeld, "Nationalist Progressivism and World War I," *Mid-America* 45 (July 1963):139–156; William E. Leuchtenburg, "Progressivism and Imperialism: The Progressive Movement and American Foreign Policy, 1898–1916," *Mississippi Valley Historical Review* 39 (December 1952): 483–504; Arthur S. Link, *Woodrow Wilson and the Progressive Era, 1910–1917* (New York: Harper and Row, 1954) and Arthur Link's multivolume study, *Wilson: The Struggle for Neutrality, 1914–1915, Wilson: Confusion and Crises, 1915–1916,* and *Wilson: Campaigns for Progressivism and Peace, 1916–1917* (Princeton, N.J.: Princeton University Press, 1960–1965); Herbert F. Margulies, "Recent Opinion on the Decline of the Progressive Movement," *Mid-America* 45 (October 1963): 250–268; Walter T. Trattner, "Progressivism and World War I: A Reappraisal," *Mid-America* 44 (July 1962):131–145; and Richard L. Watson, Jr., *The Development of National Power: The United States, 1900–1919* (Boston: Houghton Mifflin, 1976).

3. The scholarship on the American peace movement has just recently come of age. The most important collection of work has been done and continues to be done by a group of historians who founded the Con-

ference on Peace Research in History in 1964, arranged for the reprint series under the title, *The Garland Library of War and Peace,* starting in 1971, began publication of *Peace and Change: A Journal of Peace Research* in 1972, and, most recently, initiated and helped publish the *Biographical Dictionary of Modern Peace Leaders,* ed. Harold L. Josephson (Westport, Conn.: Greenwood Press, 1985).

A most extensive and thorough bibliographic essay on the peace movement is found in Charles Chatfield, *For Peace and Justice: Pacifism in America, 1914–1941* (Knoxville, Tenn.: The University of Tennessee Press, 1971), pp. 345–369. Another quite lengthy and useful essay is in Charles DeBenedetti, *Origins of the Modern American Peace Movement, 1915–1929* (Millwood, N.Y.: KTO Press, 1978), pp. 253–267. DeBenedetti updates his earlier essay in a brief bibliographic note in *The Peace Reform in American History* (Bloomington, Ind.: Indiana University Press, 1980), pp. 201–202. The best introduction to the bibliography and the subject matter of the peace movement is: Charles F. Howlett and Glen Zeitzer, *The American Peace Movement: History and Historiography,* AHA Pamphlet 261 (Washington, D.C.: American Historical Association, 1985).

4. There is a substantial literature on proposals for a defense or war council. For some of the better volumes, see: Paul Y. Hammond, *Organizing for Defense: The American Military Establishment in the Twentieth Century* (Princeton, N.J.: Princeton University Press, 1961), pp. 49–77; Samuel P. Huntington, *The Soldier and the State: The Theory and Politics of Civil-Military Relations* (Cambridge, Mass.: Belknap Press, 1957), pp. 260–263; William J. Breen, *Uncle Sam at Home: Civilian Mobilization, Wartime Federalism, and the Council of National Defense, 1917–1919* (Westport, Conn.: Greenwood Press, 1984); John M. Peoples, "The Genesis of the War Industries Board" (Ph.D. diss., University of California, Berkeley, 1942), pp. 6–10; and William Franklin Willoughby, *Government Organization in War Time and After: A Survey of the Federal Civil Agencies Created for the Prosecution of the War* (New York: D. Appleton & Co., 1919), pp. 9–11.

5. Of particular importance in this regard are the works of Arno J. Mayer: *Political Origins of the New Diplomacy, 1917–1918* (New York: Vintage Books, 1970); *Politics and Diplomacy of Peacemaking: Containment and Counterrevolution at Versailles, 1918–1919* (New York: Vintage Books, 1969); and *Dynamics of Counterrevolution in Europe, 1870–1956: An Analytic Framework* (New York: Harper and

Row, 1971). The various publications of Gabriel Kolko are also important in this regard, as are those of William Appleman Williams and especially, Williams, *The Tragedy of American Diplomacy* (New York: Dell Publishing, 1972). See also: N. Gordon Levin, Jr., *Woodrow Wilson and World Politics: America's Response to War and Revolution* (New York: Oxford University Press, 1968).

6. Figures from U.S. Department of Commerce, Bureau of the Census, *Historical Statistics of the United States, Colonial Times to 1970* (Washington, D.C.: U.S. Government Printing Office, 1975), pp. 224, 1124.

7. The literature on airplanes, aircraft, and aeronautics is enormous. A good introduction to the subject and a bibliography are provided in: Robert E. Bilstein, *Flight in America, 1900–1983: From the Wrights to the Astronauts* (Baltimore, Md: The Johns Hopkins University Press, 1984). See also: Edwin H. Rutkowski, *The Politics of Military Aviation Procurement, 1926–1934: A Study in the Political Assertion of Consensual Values* (Columbus, Ohio: Ohio State University Press, 1966); and Irving Brinton Holley, Jr., *Buying Aircraft: Materiel Procurement for the Army Air Forces* (Washington, D.C.: Office of the Chief of Military History, 1964).

8. Russell F. Weigley, *History of the United States Army* (New York: Macmillan, 1967), chap. 17, presents a succinct and insightful analysis of interwar planning by the army. All of Weigley's volumes on the military are indispensable reading for anyone interested in military history at its best. See also: Marvin A. Kreidberg and Merton G. Henry, *History of Military Mobilization in the United States Army, 1775–1945* (Washington, D.C.: Department of the Army, 1955), chaps. 12–16.

9. Hadley Cantril, ed., *Public Opinion, 1935–1946* (Princeton, N.J.: Princeton University Press, 1951), is especially helpful in gaining insights on public opinion about the military, preparedness, and war not only in the United States, but also abroad. See also: George H. Gallup, *The Gallup Poll: Public Opinion, 1935–1971,* 3 vols. 1935–1948 (New York: Random House, 1972), vol. 1.

10. A first-rate analysis of the navy is provided in the work of Harold and Margaret Sprout: *The Rise of American Naval Power, 1776–1918* (Princeton, N.J.: Princeton University Press, 1946) and *Toward a New Order of Sea Power: American Naval Policy and the World Scene, 1918–1922* (Princeton, N.J.: Princeton University Press, 1946). See also: George T. Davis, *A Navy Second to None: The Development of Modern*

American Naval Policy (New York: Harcourt, Brace and Co., 1940); Donald W. Mitchell, *History of the Modern American Navy: From 1883 through Pearl Harbor* (London: John Murry, 1947); George R. Clark and Carroll S. Alden, *A Short History of the United States Navy* (Philadelphia, Penn.: J. B. Lippincott, 1939); Paolo E. Coletta, ed., *American Secretaries of the Navy, 1775–1972,* 2 vols. (Annapolis, Md.: Naval Institute Press, 1980); Paolo E. Coletta, *The American Naval Heritage in Brief* (Washington, D.C.: University Press of America, 1978); Robert Greenhalgh Albion and Robert Howe Connery, *Forrestal and the Navy* (New York: Columbia University Press, 1962); and Arnold J. Rogow, *James Forrestal: A Study of Personality, Politics, and Policy* (New York: Macmillan, 1963).

11. Vincent Davis, *Postwar Defense Policy and the U.S. Navy, 1943–1946* (Chapel Hill, N.C.: University of North Carolina Press, 1962); Perry McCoy Smith, *The Air Force Plans for Peace, 1943–1945* (Baltimore, Md.: Johns Hopkins University Press, 1970); and Michael S. Sherry, *Preparing for the Next War: American Plans for Postwar Defense, 1941–1945* (New Haven, Conn.: Yale University Press, 1977).

12. This very sensitive theme is analyzed with fine insight based on excellent documents by Ronald Schaffer, *Wings of Judgment: American Bombing in World War II* (New York: Oxford University Press, 1985). The larger issue of the so-called strategy intellectuals who got their start during World War II is the topic of Gregg Herken, *Counsels of War* (New York: Alfred A. Knopf, 1985).

Chapter 4
The Militarization of Europe, 1914–1945

I would like to thank Dr. Auslander and Dr. Domansky for their extensive comments and helpful suggestions and discussions and John Gillis for patience.

1. The study of militarism remains tied to its early modern European origins despite efforts to expand the scope and perspective of analysis. Volker Berghahn, ed., *Militarismus* (Cologne: Kiepenheur & Witsch, 1975) and his *Militarism: The History of an International Debate, 1861–1979* (Leamington Spa: Berg Publishers, 1981); Wilfried von Bredow, *Moderner Militarismus: Analyse und Kritik* (Stuttgart: Kohlhammer, 1983). This is quite a remarkable state, if we consider that

"militarism" is a neologism of the 1860s that explicitely addressed novel phenomena of militant mass mobilization and state-building and juxtaposed them to an earlier "paternalist" (bourgeois) or aristocratic pattern of state formation.

2. The current revival of this assessment is most clearly expressed in a new wave of "apocalyptic literature." Among the most accessible and succinct statements are Edward P. Thompson and Dan Smith, eds., *Protest and Survive* (Harmondsworth: Penguin Books, 1980), Edward P. Thompson, ed., *Exterminism and Cold War* (London: Verso, 1982) and his *The Heavy Dancers: Writings on War, Past and Future* (New York: Pantheon Books, 1985). As an example of the more "apocalyptic" sentiment, see Anton Guha, *Ende. Tagebuch aus dem 3. Weltkrieg* (Kronheim/Ts.: Athenäum, 1983) or Victor Werner, *La grande peur: La 3e guerre mondiale* (Brussels: Rossel, 1976).

3. Berghahn, *Militarism,* 7–30. Werner Conze, Michael Geyer, Reinhard Stumpf, "Militarismus," *Geschichtliche Grundbegriffe,* 6 vols. ed. by Otto Brunner et al. (Stuttgart: Klett-Cotta, 1978), 4:1–48.

4. See the fascinating study by Paul Boyer, *By the Bomb's Early Light: American Thought and Culture at the Dawn of the Atomic Age* (New York: Pantheon Books, 1985). On the Soviet Union, see David Holloway, "War, Militarism, and the Soviet State," *Protest and Survive,* ed. E. P. Thompson, 129–169; Egbert Jahn, *Kommunismus—und was dann? Zur Bürokratisierung und Militarisierung des Systems der Nationalstaaten* (Reinbek: Rowohlt, 1974)

5. Emilio Willems, *A Way of Life and Death: Three Centuries of Prussian-German Militarism; an Anthropological Approach* (Nashville, Tenn.: Vanderbilt University Press, 1986) is an indication of the continuity of the argument.

6. Joseph A. Schumpeter, "Soziologie der Imperialismen," *Archiv für Sozialwissenschaft und Sozialpolitik* 46 (1919):1–39, 275–310; Herbert Spencer, *Principles of Sociology,* 2 vols. (repr., Westport, Conn.: Greenwood Press, 1975), vol. 2. However, it should be noted that in setting "militant" and "industrial" societies apart, Spencer pointed to a process of "re-barbarization" in the late nineteenth century that is exactly our concern. David Wiltshire, *The Social and Political Thought of Herbert Spencer* (Oxford: Oxford University Press, 1978).

7. *Denn die Liebe kann nicht um vieles jünger sein als die Mordlust:* This grand theme of eighteenth and nineteenth century literature (danger

and desire/sexuality and death) has entered the scientific canon through Sigmund Freud. See in particular his "Thoughts for the Time on War and Death," *Standard Edition of the Complete Works of Sigmund Freud,* 24 vols. ed. by James Strachey (London, Hogarth, 1975) 14:275–302, or "Why War?", ibid., 22: 197–215. For the context, see Sherry B. Ortner, "Is Female to Male as Nature Is to Culture?" in *Woman, Culture, and Society,* ed. Michelle Z. Rosaldo and Louise Lamphere (Stanford: Stanford University Press, 1974), pp. 67–87. The whole problem, of course, has been traditionally illustrated (Roswitha Flatz, *Krieg im Frieden. Das aktuelle Militärstück auf dem Theater des deutschen Kaiserreiches* [Frankfurt: Klostermann, 1976]) by the history of uniforms and their implicit tension between orderliness/subordination—flamboyance/seduction, which is oddly repressed by Nathan Joseph, *Uniforms and Non-Uniforms: Communication through Clothing* (Westport, Conn.: Greenwood Press, 1986). This latter contribution is a good indication of the "purification" of once "potent" ideas in the late twentieth century.

8. Max Scheler, "Über Gesinnungs- und Zweckmilitarismus: Eine Studie zur Psychologie des Militarismus," in *Krieg und Aufbau,* ed. Max Scheler (Leipzig: Verlag der weiβen Bücher, 1916), pp. 167–195. Franz Carl Endres, "Militarismus als Geistesverfassung des Nichtmilitärs," in *Militarismus,* ed. V. Berghahn, pp. 99–101.

9. The definite statement of this position is Alfred Vagts, *A History of Militarism: Civilian and Military,* rev. ed. (New York: Free Press, 1959).

10. Samuel P. Huntington, *The Soldier and the State: The Theory and Politics of Civil-Military Relations* (New York: Vintage Books, 1957); Samuel E. Finer, *The Man on Horseback: The Role of the Military in Politics* (London: Pall Mall Press, 1962).

11. Gordon A. Craig, *The Politics of the Prussian Army, 1640–1945* (London: Oxford University Press, 1955); Hans Rosenberg, *Bureaucracy, Aristocracy, and Autocracy: The Prussian Experience, 1660–1815* (Boston: Beacon Press, 1966).

12. Morris Janowitz, *The Professional Soldier: A Social and Political Portrait,* reprint (New York: Free Press, 1971), but see John Wheeler-Bennett, *The Nemesis of Power: The German Army in Politics 1918–1945* (London: Macmillan, 1953).

13. Hans-Ulrich Wehler, *The German Empire, 1871–1918* (Leamington Spa: Berg, 1985); W. Sauer, "Die politische Geschichte der deutschen

Armee und das Problem des Militarismus," *Politische Vierteljahresschrift* 6 (1965):341–353; *idem.*, "National Socialism: Totalitarianism or Fascism," *American Historical Review* 73 (1967): 404–424.

14. On the debates about the German military heritage in the 1950s, see Donald Abenheim, *A Valid Heritage: West German Rearmament and the Debate over Germany's Military Tradition* (Princeton: Princeton University Press, forthcoming); Manfred Messerschmidt, "Das Verhältnis von Wehrmacht und NS Staat und die Frage der Traditionsbildung," in *Tradition als Last: Legitimationsprobleme der Bundeswehr*, ed. Klaus M. Kodalle (Cologne: Verlag Wissenschaft und Politik, 1981).

15. D. C. Rapoport, "Military and Civil Societies: The Contemporary Significance of a Traditional Subject in Political Theory," *Political Studies* 12 (1984):178–201. The eighteenth-century paradigm of a segmentation between military and civil society is at the core of almost every study on "civil-military relations." These studies dutifully accept an intellectual framework that simply does not fit twentieth-century conditions.

16. The classic statement on this phenomenon is Harold D. Lasswell, "The Garrison State and the Specialists on Violence," *American Journal of Sociology* 47 (1941):455–468. See also Donald C. Watt, *Too Serious a Business: European Armed Forces and the Approach to the Second World War* (London: Temple Smith, 1975) and Brian Bond, *War and Society in Europe 1870–1970* (Oxford: Oxford University Press, 1986).

17. MacGregor Knox, "Conquest, Foreign and Domestic, in Fascists Italy and Nazi Germany," *Journal of Modern History* 56 (1984):1–57. Michael Geyer, "Krieg als Sozialpolitik: Anmerkungen zu neueren Arbeiten über das Dritte Reich im Zweiten Weltkrieg," *Archiv für Sozialgeschichte* 26 (1986):557–601.

18. Charles Bright and Michael Geyer, "For a Unified History of the World in the Twentieth Century," *Radical History Review* 39 (1987): 69–91.

19. See the essays by John Shy, Peter Paret, Gunther E. Rothenberg, Douglas Porch, and Michael Geyer in Peter Paret, ed., *Makers of Modern Strategy from Machiavelli to the Nuclear Age* (Princeton: Princeton University Press, 1986).

20. This is at the core of the European-wide "short-war illusion." It would be entirely unwarranted to limit this notion to Germany and to explain it in reference to either geopolitical expediences or the instabilities of German society. See the various essays in Paret, *Makers of Modern Strategy* and Jack Snyder, *The Ideology of the Offensive: Military Deci-*

sion Making and the Disaster of 1914 (Ithaca: Cornell University Press, 1984); L. L. Farrar, *The Short-War-Illusion; German Policy, Strategy and Domestic Affairs August-December 1914* (Santa Barbara: ABC-Clio, 1973).

21. Bond, *War and Society,* pp. 72–99; Roland N. Stromberg, *Redemption by War: The Intellectuals and 1914* (Lawrence: Regents Press of Kansas, 1982). The best analyses of this kind are Roger Chickering, *We Men who Feel Most German: A Cultural Study of the Pan-German League 1886–1914* (Boston: Allen & Unwin, 1984); Anne Summers, "Militarism in Great Britain before the Great War," *History Workshop Journal* 2 (1976):104–123; John Springhall, *Youth, Empire, and Society: British Youth Movement, 1893–1940* (London: Croom Helm, 1977). A similar analysis for France is missing, though France (and Italy) may indeed prove to be the best cases. See the suggestions in Claude Quiguer, *Femmes et machines de 1900: lecture d'une obsession Modern Style* (Paris: Klincksieck, 1979).

22. This is the gist of a new interpretation of the outbreak of World War I by Samuel Williamson, "Vienna and July 1914: The Origins of the Great War once again," *Essays on World War I: Origins and Prisoners of War,* ed. Samuel Williamson and Peter Pastor (New York: Brooklyn College Press, 1983), pp. 9–36, but also see Vladimir Dedijer: *The Road to Sarajevo* (New York: Simon and Schuster, 1966). See also Joachim Remak, "1914—The Third Balkan War," *The Origins of the First World War; Great Power Rivalry and German War Aims,* ed. H. W. Koch (London: Macmillan, 1984).

23. The tendency to exclude the political mainstream as the main support for war is again European-wide. Hence everyone scrambles to explain why this mainstream supported the war unanimously and enthusiastically once it had started. The fact, however, is that the political center in all European countries rallied around rearmament before the war and began to establish a new political identity on this basis. Gerd Krumeich, *Aufrüstung und Innenpolitik in Frankreich vor dem Ersten Weltkrieg: Die Einführung der dreijährigen Dienstpflicht 1913–1914* (Wiesbaden: Steiner, 1980); Robert Scally, *The Origins of the Lloyd George Coalition: The Politics of Social Imperialism 1900–1918* (Princeton: Princeton University Press, 1975) or John Grigg, *Lloyd George: The People's Champion, 1902–1911* (Berkeley and Los Angeles: University of California Press, 1978). In this sense, we can, indeed, speak of a rising "Cold War" tide, as opposed to Volker Berghahn,

Rüstung und Machtpolitik. Zur Anatomie des "Kalten Krieges" vor 1914 (Düsseldorf: Droste, 1973). The main opposition against this view derives from a nineteenth-century liberal bias, which cannot imagine bourgeois politics as the source of war. Michael Howard, *War and Liberal Conscience* (New Brunswick: Rutgers University Press, 1978).

24. E. Carr, *The Twenty Years' Crisis, 1919–1939,* reprint (New York: Harper Torchbooks, 1964).

25. Hew Strachan, *European Armies and the Conduct of War* (Boston: Allen & Unwin, 1983); William McNeill, *The Pursuit of Power. Technology, Armed Force, and Society since A.D. 1000* (Chicago: University of Chicago Press, 1982), pp. 307–361.

26. Friedrich G. Jünger, *Die Perfektion der Technik,* 5th ed. (Frankfurt: Klostermann, 1968), pp. 180–197 is the most powerful and original statement of this notion. On its symbolic interpretation, see Eric J. Leed, *No Man's Land: Combat and Identity in World War I* (Cambridge: Cambridge University Press, 1979).

27. This point is most cogently argued in Elisabeth Domansky, "Der Erste Weltkrieg," in *Die bürgerliche Gesellschaft in Deutschland,* ed. Lutz Niethammer (Munich: Beck, forthcoming) as well as in *idem,* "Politische Dimensionen von Jugendprotest und Generationenkonflikt in der Zwischenkriegszeit in Deutschland," *Jugendprotest und Generationenkonflikt in Europa im 20. Jahrhundert* (Bonn: Verlag Neue Gesellschaft, 1986), pp. 113–137. See also Klaus Vondung, ed., *Kriegserlebnis: Der Erste Weltkrieg in der literarischen Gestaltung und symbolischen Deutung der Nationen* (Göttingen: Vandenhoeck & Ruprecht, 1980).

28. This is the main theme of the political history of World War I. Gerald Feldman, *Army and Industry and Labor, 1914–1918* (Princeton: Princeton University Press, 1966); John F. Godfrey, *Capitalism at War: Industrial Policy and Bureaucracy in France 1914–1918* (Leamington Spa: Berg, 1987); Kathleen Burk, ed., *War and the State: The Transformation of British Government, 1914–1919* (Boston: Allen & Unwin, 1982).

29. This seems to me the historical source for Habermas's segregation of *Systemwelt* and *Lebenswelt,* which has a venerable albeit unacknowledged tradition. In German parlance this phenomenon was linked to a purported crisis of "liberalism." See Alfred Weber, *Die Krise des modernen Staatsgedankens in Europa* (Stuttgart: Deutsche Verlags-Anstalt, 1925); Carl Schmitt, *The Crisis of Parliamentary Democracy* (Cambridge: MIT Press, 1985). In France it was linked to the republican oppo-

sition against the *étatism industriel,* see R. Carnot, *L'Etatisme industriel* (Paris: Payot, 1920); and for the bureaucratic centralization of the state see Leon Duguit, *Law in the Modern State,* trans. F. and H. Laski (New York: B. W. Huebsch, 1919).

30. Eugen Weber, *Peasants into Frenchmen: The Modernization of Rural France, 1870–1914* (Stanford: Stanford University Press, 1976) develops this point most forcefully. The same theme is a subtext in David Blackbourn and Geoff Eley, *The Peculiarities of German History: Bourgeois Society in Nineteenth-Century Germany* (New York: Oxford University Press, 1984). One may note, on the other hand, the striking degree of the nationalization of politics in Great Britain before the war. The common problem of continental European studies is their tendency to locate the process of nationalization in the state. It rather seems that parastatal agencies became the main agents of centralization in the twentieth century. For the study of the relation between organized labor and industry in the context of "organized capitalism," see Charles Maier, *Recasting Bourgeois Europe: Stabilization in France, Germany, and Italy in the Decade after World War I* (Princeton: Princeton University Press, 1975); Heinrich-August Winkler, ed., *Organisierter Kapitalismus: Voraussetzungen und Anfänge* (Göttingen: Vandenhoek & Ruprecht, 1974), the study of interest groups, see Suzanne Berger, *Organizing Interests in Western Europe: Pluralism, Corporatism, and the Transformation of Politics* (Cambridge: Cambridge University Press, 1981). On the very "thick" parastatal mediation between "state" and "society" in the welfare sector see Michael Geyer, "Ein Vorbote des Wohlfahrtsstaates. Die Kriegsopferversorgung in Frankreich, Deutschland und Großbritannien nach dem Ersten Weltkrieg," *Geschichte und Gesellschaft* 9 (1983): 230–277.

31. Derek Aldcroft, *From Versailles to Wall Street: The International Economy in the 1920s* (Berkeley and Los Angeles: University of California Press, 1977); Arno J. Mayer, *Politics and Diplomacy of Peacemaking: Containment and Counterrevolution at Versailles 1918–1919* (New York: Alfred A. Knopf, 1968).

32. Christopher M. Andrew, A. S. Kanya-Forstner, *France Overseas: The Great War and the Climax of French Imperial Expansion* (London: Thames and Hudson, 1981); See especially the argument in Paul Kennedy, *The Rise and Fall of the Great Powers* (New York: Random House, 1987).

33. This is, of course, an argument that fits the eighteenth- and

nineteenth-century tradition of militarism; see for example, Gerhard Ritter, *The Sword and the Scepter: The Problem of Militarism in Germany,* 4 vols. (Coral Gables, Fla.: University of Miami Press, 1969–1973), esp. vols. 3 and 4, but also Martin Kitchen, *The Silent Dictatorship: The Politics of the German High Command under Hindenburg and Ludendorff, 1916–1918* (London: Oxford University Press, 1976) or Jere C. King, *Generals and Politicians; Conflict between France's High Command, Parliament, and Government, 1914–1918* (Berkeley and Los Angeles: University of California Press, 1951).

34. Leed, *No Man's Land,* pp. 115–162.

35. This is, unfortunately, the main metaphor for the current wave of "war and society" studies, as, for example, Arthur Marwick, *War and Social Change in the Twentieth Century: A Comparative Study of Great Britain, France, Germany, Russia, and the United States* (London: Macmillan, 1974). While these studies expand the scope of military history (and, conversely, social history), they mostly tend to externalize war and reify it into a barely veiled metahistorical actor. Neither suffices; the common, conceptual framework for "war and society" studies sees the encompassing nature of war, but is unable to explain war as a violent mode of social organization in the twentieth century.

36. F. G. Jünger, *Perfektion der Technik,* points out very nicely the simultaneous increase in institutional domination and lack of "perspective." It is this the *"pilotage à vue"* (tunnel vision) that is often considered to be a general feature of the twentieth century state.

37. Michael Geyer, "The State in National Socialist Germany," in *Statemaking and Social Movements: Essays in History and Theory,* ed. Charles Bright and Susan Harding (Ann Arbor: University of Michigan Press, 1984), pp. 193–232.

38. In this respect, I follow McNeill, *The Pursuit of Power* and the pathbreaking study of Raymond Aron, *The Century of Total War* (Boston: Beacon Press, 1955) who both argue for a new "stage" in the development of the "forces of destruction." However, I would consider neither "technology" nor "managerial capabilities" as the driving force of this process. Rather, these forces are consequences of the competitive organization of national and international processes of the social formation. Hence, militarization is the product of interactions and it requires choices in shaping them.

39. Allan S. Milward, *War, Economy, and Society, 1939–1945* (Berkeley and Los Angeles: University of California Press, 1977).

40. This definition is purposefully broad; for militarized expansion as exemplified by National Socialism and Japanese militarism is only one variant of militarism in the twentieth century. Twentieth-century imperial formations—the politics of forced colonial "development"—as well as the Pax Americana and Pax Sovietica after 1945 are quite fundamentally different, but they are "militarist" social formations according to this definition. A more elaborate discussion about the global expansion of "militarism" in World War II must wait for the moment.

41. The newest in a long series of studies is J. F. Godfrey, *Capitalism at War;* see also the discussion in Richard Kuisel, *Capitalism and the State in Modern France* (Cambridge: Cambridge University Press, 1981), pp. 31–58; Patrik Fridenson, *Histoire des usines Renault: naissance de la grand entreprise, 1880–1939* (Paris: Seuil, 1970); Gerd Hardach, "La mobilization industrielle au 1914–1918: production, planification, et ideologie," in *1914–1918; L'autre front,* ed. Patrik Fridenson (Paris: Les editions ouvriers, 1977).

42. See the older study by Pierre Larigaldie, *La politique économique interalliee: Les organismes interalliees de control économique* (Paris: Longin, 1925); J. A. Salter, *Allied Shipping Control: An Experiment in International Organization* (Oxford: Clarendon Press, 1921); Ellen Schrecker, *The Hired Money: The French Debt to the United States, 1917–1929* (New York: Arno, 1978); Kathleen Burk, *Britain, America, and the Sinews of War, 1914–1918* (Boston: Allen & Unwin, 1985).

43. It fits this pattern well that war-related production increased tremendously and seems to have outpaced British armaments; see Ministère de la guerre, état-major des armées, service historique de l'armée, *Les armées françaises dans la grande guerre* (Paris, Imprimerie Nationale, 1937) 9:567–571, on ammunition. Generally, see C. Reboul, *Mobilization industrielle, Les fabrications de guerre en France de 1914 à 1918* (Nancy: Berger-Levault, 1925), vol. 1; and it is of equal importance that the American expeditionary forces were armed in France. Both amount to an export of war-related production to France, while Great Britain and the United States controlled the resource bases of production. On the struggle over control, see especially Marc Trachtenberg, "A New Economic Order: Etienne Clementel and French Economic Diplomacy during the First World War," *French Historical Studies* 10 (1977): 315–341. French economic history is too impressed by the transformation of industry and too little concerned with the internationalization of the relations of production.

44. Andrew and Kanya-Forster, *France Overseas;* M. Michel, "Un 'mythe': la 'Force Noir' avant 1914," *Relations Internationales* 1 (1974): 83–90 and *idem,* "Le recrutement du triailleurs en A.O.F. pendant la première guerre mondiale; essai de bilan statistique," *Revue française d'histoire d'outre-mer* 60 (1973):644–660.

45. Michel Huber, *La population de la France pendant la guerre* (Paris, New Haven: Yale University Press; Les presses universitaries de France, 1931), pp. 94–103, 198–205, p. 426 (for impact on the countryside); see also *Les Armées françaises,* vol. 11 and B. Nogaro and L. Weil, *La main-d'oeuvre étrangère et colonials pendant la guerre* (Paris: Presses Universitaries du France, 1926) on the role of colonial labor, and Antoine Prost, *Les anciens combattants et la société française,* 3 vols. (Paris: Presses de la Fondation Nationale des Sciences Politiques, 1977), for regional patterns of mobilization. Unfortunately, the mobilization of the countryside is less well researched than one might expect.

46. Steven C. Hause, "More Minerva than Mars: The French Women's Rights Campaign and the First World War," in *Behind the Lines: Gender and the Two World Wars,* ed. Margaret Randolph Higonnet and Jane Jenson (New Haven: Yale University Press, 1987), pp. 99–113.

47. Max Gallo, "Quelques aspects de la mentalité et des comportement ourvriers dans les usines de guerre," *Le mouvement social* 56 (1966):3–34; Jean-Jaques Becker, *The Great War and the French People* (New York: St. Martin's Press, 1986).

48. Becker, *The Great War,* esp. pp 113–131, with the emphasis on the stolid acceptance of the war in the countryside. A much more activist picture is drawn by Prost, *Anciens Combattants,* who points to the reconsolidation of social relations.

49. Antoine Prost, "Combattants et politiciens: Le discours mythologique sur la politique entre de deux guerres," *Le Mouvement social* 85 (1973):117–154.

50. Marc Trachtenberg, *Reparations in World Politics: France and European Economic Diplomacy 1916–1923* (New York: Columbia University Press, 1980).

51. Bernard Porter, *Britain, Europe and the World, 1850–1982: Delusions of Grandeur* (Boston; Allen & Unwin, 1983); C. Ernest Fayle, *Seaborne Trade (History of the Great War, by Direction of the Historical Section of the Committee of Imperial Defense),* 3 vols. (London: Murray, 1920–1924); and the essays in *The Cambridge History of the British*

Empire, ed. Ernest A. Benians et al., 8 vols. (Cambridge: Cambridge University Press, 1932–1936) vols. 5–8.

52. Aldcroft, *From Versailles to Wall Street*, chap. 1; Alan S. Milward, *The Impact of the World Wars on the British Economy* (London: Macmillan, 1970); League of Nations, *Memorandum on Currency and Central Banks, 1913–1924*, 2 vols. (Geneva: League of Nations, 1925); *idem, Memorandum on Production and Trade*, (Geneva, League of Nations, 1926).

53. *India's Contribution to the Great War*, published by Authority of the Government of India (Calcutta: Superintendent Government Printing, 1923); C. Ellinwood DeWitt and S. D. Pradhan, eds. *India and World War I* (New Delhi: Manohar, 1978); B. R. Tomlinson, *The Political Economy of the Raj, 1914–1947: The Economics of Decolonization* (London: Macmillan, 1979); Dietmar Rothermund, *Die politische Willensbildung in Indien, 1900–1960* (Wiesbaden: Harrassowitz, 1965).

54. Australia: Leslie L. Robson, *The First AIF: A Study of recruitment 1914–1918* (Melbourne: Melbourne University Press, 1970); Jane Ross, *The Myth of the Digger: The Australian Soldier in Two Wars* (Sydney: Hale & Iremonger, 1985); Canada: Desmond Morton, *A Peculiar Kind of Politics: Canada's Overseas Ministry in the First World War* (Toronto: University of Toronto Press, 1982); J. L. Granatstein and J. M. Hitsman, *Broken Promises: A History of Conscription in Canada* (Toronto: Oxford University Press, 1977). A good study on South Africa is missing. On the impact of the war generally see Ian M. Drummond, *Imperial Economic Policy, 1917–1939: Studies in Expansion and Protection* (London: Allen & Unwin, 1974).

55. In comparative perspective, see Dick Geary, *European Labor Protest, 1848–1939* (New York: St. Martin's Press, 1981); Gerry R. Rubin, *War, Law, and Labor: The Munitions Act, State Regulation, and the Unions, 1915–1920* (Oxford: Oxford University Press, 1987); R. Klepsch, *British Labour im Ersten Weltkrieg: Die Ausnahmesituation des Krieges 1914–1918 als Problem und Chance der britischen Arbeiterbewegung* (Göttingen: T. Bantz, 1983).

56. Jay M. Winter, *Socialism and the Challenge of War* (London: Routlege & Kegan Paul, 1974); *idem*, "Military Fitness and Public Health in Britain in the First World War," *Journal of Contemporary History* 15 (1980):211–244; P. Dewey, "Military Recruiting and the British Labour Force," *Historical Journal* 27 (1984):199–223; Gail

Braybon, *Women Workers and the First World War: The British Experience* (London: Croom Helm, 1981).

57. Jay M. Winter, *War and the British People* (Cambridge, Mass.: Harvard University Press, 1986), p. 92.

58. Kenneth O. Morgan, *Consensus and Disunity: The Lloyd George Coalition Government, 1918–1922* (New York: Oxford University Press, 1979); Paul B. Johnson, *Land Fit for Heroes* (Chicago: University of Chicago Press, 1968); Stanley Shapiro, "The Great War and Reform: Liberals and Labour 1917–1919," *Labor History* 12 (1971):323–344.

59. The literature on the opposition against the war in Great Britain is endless, but it tends to overlook the fact that the various groups were neither comrades nor sisters, as Richard J. Evans, *Comrades and Sisters: Feminism, Socialism, and Pacifism in Europe, 1870–1945* (New York: St. Martin's Press, 1987) suggests. See also the summary in Francis L. Carsten, *War against the War: British and German Radical Movements in the First World War* (Berkeley and Los Angeles: University of California Press, 1982). The same cannot be said about the various prowar groups. Partly they are less well studied, partly there was no prowar social movement—though there was prowar sentiment—as in France and Germany. The fate of the Silver Badge Party as opposed to the veterans movements on the continent is indicative. Stephen R. Ward, "Great Britain: Land Fit for Heroes Lost," *The War Generation. Veterans of the First World War* (Port Washington, N.Y.: Kennikat Press, 1975), pp. 10–37; *idem*, "The British Veterans' Ticket of 1918," *Journal of British Studies* 8 (1968):155–169.

60. Margaret L. Barnett, *British Food Policy during the First World War* (Boston: Allen & Unwin, 1985); Deborah Dwork, *War Is Good for Babies and Other Young Children: a History of the Infant and Child-welfare Movement in England* (London: Tavistock Press, 1987).

61. Geyer, "Vorbote des Wohlfahrtsstaates," pp. 258–264.

62. Elements of ethnic politics were as present in all the belligerent countries as gender and class politics, but they were successfully externalized and subordinated in the imperial countries. On the centrality of this aspect in Austria-Hungary, see Richard Georg Plaschka et al., *Innere Front: Militärassistenz, Widerstand und Umsturz in der Donaumonarchie 1918*, 2 vols. (Vienna: Verlag fur Wissenschaft und Politik, 1974); Wiktor Sukiennicki, *East Central Europe during World War I: From Foreign Domination to National Independence* (Boulder, Colo.: Eastern European Monographs, 1984).

63. The first study to point to this fact was Feldman, *Army, Industry and Labor.*

64. This is the main theme of the German debate on "corporatism": Winkler, *Organisierter Kapitalismus.* The role of parliament is much debated: Erich Matthias, ed., *Der Interfraktionelle Ausschuss 1917/1918* (Düsseldorf: Droste, 1959); Reinhard Schiffers and Manfred Koch, eds., *Der Hauptausschuss des deutschen Reichstages, 1915–1918* (Düsseldorf: Droste, 1981–82); Udo Bermbach, *Vorformen parlamentarischer Kabinettsbildung in Deutschland: Der Interfraktionelle Ausschuss 1917/1918* (Cologne: Westdeutscher Verlag, 1967). But this role is misunderstood: parliament gained importance not as a policy-setting institution; rather it became a central mediator in organizing domination—which may also help to explain its persistent delegitimation in the 1920s. On the power of parastatal mediating agencies see, among others, Gunther Mai, *Kriegswirtschaft und Arbeiterbewegung in Württemberg 1914–1918* (Stuttgart: Klett-Cotta, 1983); Herman Schäffer, *Regionale Wirtschaftspolitik in der Kriegswirtschaft. Staat, Industrie und Verbände während des Ersten Weltkrieges in Baden.* (Stuttgart: Kohlhammer, 1983). On welfare services, see the forthcoming thesis of Young Sun Hong, "From Disciplinary Charity to the Modern Social Welfare System: The Politics of Welfare Reform in Germany 1880–1933" (Ph.D. diss., University of Michigan, 1988).

65. Hans-J. Bieber, *Gewerkschaften in Krieg und Revolution. Arbeiterbewegung, Industrie, Staat und Militär in Deutschland 1914–1920* (Hamburg: Christians, 1981); G. Feldman, "German Business between War and Revolution: The Origins of the Stinnes-Legien Agreement," in *Entstehung und Wandel der modernen Gesellschaft. Festschrift für Hans Rosenberg,* by Gerhard A. Ritter (Göttingen: Vandenhoeck & Ruprecht, 1970), pp. 312–341; Dirk H. Müller, "Gewerkschaften, Arbeiterausschüsse und Arbeiterrräte in der Berliner Kriegsindustrie 1914–1918," in *Arbeiterschaft in Deutschland 1914–1918. Studien zu Arbeitskampf und Arbeitsmarkt im Ersten Weltkrieg,* ed. Gunther Mai (Düsseldorf: Droste, 1985). Marc Ferro, *The Great War, 1914–1918* (London: Routlege & Kegan Paul, 1973), pp. 147–179 discusses this point succinctly.

66. The process of depoliticized and institutionally organized participation characterizes the German "way" of militarization. National differences in this respect are most pronounced in the mobilization of women. Karin Hausen, "The German Nation's Obligation to the Heroes' Widows in World War I," in *Behind the Lines,* pp. 126–140; Christoph

Sachsse, *Mütterlichkeit als Beruf: Sozialarbeit, Sozialreform und Frauenbewegung, 1917–1929* (Frankfurt: Suhrkamp, 1986); Ursula von Gerstorff, *Frauen im Kriegsdienst* (Stuttgart: Deutsche Verlags-Anstalt, 1969); Ute Frevert, *Frauen-Geschichte: Zwischen Bürgerlicher Verbesserung und Neuer Weiblichkeit* (Frankfurt: Suhrkamp, 1986). These studies all point to the high degree of nationalization. There is no good study of food riots that would show the other "politicized" side of mobilization, and, as one comes to expect, the role of women in the German revolution is completely unexplored.

67. Rudolph Meerwarth et al., *Die Einwirkungen des Kriegs aud Bevölkerungsbewegung, Einkommen und Lebenshaltung in Deutschland* (Stuttgart: Deutsche Verlags-Anstalt, 1932); Albrecht Mendelssohn-Bartholdy, *The War and German Society: The Testament of a Liberal* (New Haven: Yale University Press, 1937); Jürgen Kocka, *Facing Total War: German Society 1914–1918* (Cambridge, Mass.: Harvard University Press, 1984) and the subtle critique of Gottfried Schramm, "Klassengegensätze im Ersten Weltkrieg: Zu Jürgen Kockas Gesellschaftsmodel," *Geschichte und Gesellschaft* 2 (1976):244–260.

68. Gottfried Schramm, "Militarisierung und Demokratisierung: Typen der Massenintegration im Ersten Weltkrieg," *Francia* 3 (1975): 476–497; Carsten, *War against War*. On the right radical mobilization, see Abraham J. Peck, *Radicals and Reactionaries: The Crisis of Conservatism in Wilhelmine Germany* (Washington, D.C.: University Press of America, 1969); Dirk Stegman, *Die Erben Bismarcks: Parteien und Verbände in der Spätphase des Wilhelminischen Deutschlands* (Cologne: Kiepenheuer & Witsch, 1970). On the pro- and antiwar mobilization: James M. Diehl, "The Organization of German Veterans, 1917–1919," *Archiv für Sozialgeschichte* 11 (1971):335–348. On the symbolic dimensions of mobilization: Klaus Vondung, "Geschichte als Weltgericht: Genesis und degradation einer Symbolik," *Kriegserlebnis,* 62–84. There is nothing on the symbolism of the antiwar movement.

69. C. Bertrand, ed., *Revolutionary Situations in Europe, 1917–1922: Germany, Italy, Austria-Hungary* (Montreal: Centre Interuniversitaire des etudes européennes, 1977).

70. Guy Petroncini, *Les mutinieries de 1917* (Paris: Presses Universitaires de France, 1967); Piero Melograni, *Storia politica della grande guerre, 1915–1918* (Bari: Laterza, 1969); Alberto Monticone, *Gli Italiani in uniforme, 1915–1918* (Bari: Laterza, 1972); W. B. Lincoln, *Passage through Armageddon: The Russians in War and Revolution, 1914–1918*

(New York: Simon and Schuster, 1986); Bushnell, "Peasants in Uniform: The Tsarist Army as Peasant Society," *Journal of Social History* 13 (1979/80):565–576.

71. Noncooperation and the ritualization of violence was of course a general response to the ardors of industrial warfare that needs further research. John Ashworth, *Trench Warfare, 1914–1918: The Live and Let Live System* (New York: Holmes & Meier, 1980).

72. Prost, *Anciens Combattants,* vol 2.

73. In addition to Melograni, *Storia politica,* see Enrico Forcella and Alberto Monticone, eds., *Plotone di esecuzione: I processi della prima guerra mondiale* (Bari: Laterza, 1986) and the older study by Arrigo Serpieri, *La guerra e le classi rurali italiane* (Bari: Laterza, 1930).

74. Allan K. Wildman, *The End of the Russian Imperial Army: The Old Army and the Soldiers' Revolt, March–April 1917* (Princeton: Princeton University Press, 1980).

75. Michael Geyer, *Deutsche Rüstungspolitik, 1860–1980* (Frankfurt: Suhrkamp, 1984), pp. 115–121; Ulrich Kluge, *Soldatenräte und Revolution: Studien zur Militärpolitik in Deutschland 1918/1919* (Göttingen: Vandenhoeck & Ruprecht, 1975).

76. Ulrich Kluge, *Die deutsche Revolution 1918–1919: Staat, Politik und Gesellschaft zwischen Weltkrieg und Kapp-Putsch* (Frankfurt: Suhrkamp, 1985). The competition with the Right is commonly underestimated. So far there is no comprehensive study on right-wing association-building, paramilitary mobilization, and the reestablishment of bourgeois order with the help of militias (*Einwohnerwehren*). James M. Diehl, *Paramilitary Politics in Weimar Germany* (Bloomington: Indiana University Press, 1977) is a good start.

77. Domansky, "Der Erste Weltkrieg," forcefully points to this conclusion. James M. Diehl, "Veterans' Politics under Three Flags," in *The War Generation,* 135–189. Wolfram Wette and Karl Holl, eds., *Pazifismus in der Weimarer Republik: Beiträge zur historischen Friedensforschung* (Paderborn: Schöningh, 1981) and Robert W. Whalen, *Bitter Wounds: German Victims of the Great War, 1914–1939* (Ithaca: Cornell University Press, 1984), show the power of the antiwar sentiment, but they are mute about the agents of this process as well as about the rhetoric of mobilization.

78. This argument is summarized in McNeill, *The Pursuit of Power,* pp. 307–344.

79. It is common to refer to the lack of preparedness of France and

especially of great Britain, compared to Germany. However, studies on the interwar years tend to overlook that despite demobilization and drastic cuts in the military budgets (especially in Great Britain), the number of men under arms in Europe (not to speak of the colonies) and military expenditures, even in the narrow sense of the word, never dipped below the 1913 level. If one adds up all war-related expenditures (pensions and so forth), one begins to appreciate fully the degree of militarization in Europe after 1918. Nokhim M. Sloutzky, *World Armaments Race* (Geneva: Geneva Research Center, 1941).

80. Francis L. C. Lyons, *Internationalism in Europe, 1815–1914* (Leyden: A. W. Sythoff, 1963); Geoffrey Best, *Humanity in Warfare* (New York: Columbia University Press, 1980); Michael Howard, ed., *Restraints on War: Studies in the Limitation of Armed Conflict* (Oxford: Oxford University Press, 1979); Anthony P. Adamthwaite, *The Lost Peace: International Relations in Europe 1918–1939* (New York: St. Martin's Press, 1981).

81. This argument poses some problems in regard to the discussion on the German *Sonderweg*. I follow Geoff Eley, "Army, State, and Civil Society: Revisiting the Problem of German Militarism," in his *From Unification to Nazism: Reinterpreting the German Past* (Boston: Allen & Unwin, 1986), pp. 85–109, but would emphasize more strongly the transition from an old state- and caste-based "militarism" to the cult of violence by militarist movements. First steps in this direction are made by Stig Förster, "Alter und Neuer Militarismus im Kaiserreich," in *Bereit zum Krieg: Kriegsmentalität im wilhelminischen Deutschland, 1890–1914,* ed. Jost Dülffer and Karl Holl (Göttingen: Vandenhoeck & Ruprecht, 1986), pp. 122–145.

82. Anthony P. Adamthwaite, *France and the Coming of the Second World War, 1936–1939* (London: Cass, 1977); Gustav Schmidt, *The Politics and Economics of Appeasement: British Foreign Policy in the 1930s* (New York: St. Martin's Press, 1986).

83. The newest collections are Holger Klein, ed., *The First World War in Fiction: A Collection of Critical Essays* (New York: Barnes & Noble, 1977); Vondung, *Kriegserlebnis;* Bernd Hüppauf, ed., *Ansichten vom Krieg: Vergleichende Studien zum Ersten Weltkrieg in Literatur und Gesellschaft* (Königstein/TS: Forum Academicum, 1984).

84. The term itself is indicative: rearmament is the means for the "strengthening of the nation," the insistence on "manliness" and "effi-

ciency" *(Leistung),* and the "purification" of the German *Volk*. Especially the latter meaning linked rearmament and racialist policies. On the origins of this transferrence of symbols, see George L. Mosse, "Soldatenfriedhöfe und nationale Wiedergeburt" and Ulrich Linse, " 'Saatfrüchte sollen nicht vermahlen werden!' Resymbolisierung des soldatischen Todes," *Kriegserlebnis,* pp. 241–261, 262–274.

85. Wolfram Wette, "Ideologien, Propaganda und Innenpolitik als Voraussetzungen der Kriegspolitik des Dritten Reiches," in *Ursachen und Voraussetzungen der deutschen Kriegspolitik,* ed. Wilhelm Deist et al. (Stuttgart: Deutsche Verlags-Anstalt, 1979), pp. 25–173.

86. Paul Fussell, *The Great War and Modern Memory* (New York: Oxford University Press, 1975) and John Aldridge, *After the Lost Generation: A Critical Study of the Writers of Two Wars* (New York: Arbor House, 1985). For France: *La guerre et la paix dans les lettres françaises de la guerre du Rif a la guerre d'Espagne (1925–1939)* (Reims: Presses Universitaires de Reims, 1983).

87. Horst Boog et al., *Der Angriff auf die Sowjetunion* (Stuttgart: Deutsche Verlagsanstalt, 1983), esp. pp. 908–935.

88. See the statistical evidence in Peter Flora et al., eds., *State, Economy, and Society in Western Europe 1815–1975,* 2 vols. vol. 1: *The Growth of Mass Democracies and Welfare States* (Frankfurt, London, Chicago: Campus, Macmillan, St. James Press, 1983).

89. On the "cult of violence," see Ernst Nolte, *Three Faces of Fascism: Action Française, Italian Fascism, National Socialism* (New York: Holt, Rinehart, Winston, 1966); Klaus Theweleit, *Male Fantasies* (Minneapolis: University of Minnesota Press, 1987). More broadly on the literary reflection of militarized identities, see Waltraud Amberger, *Männer, Krieger, Abenteurer: Der Entwurf des soldatischen Mannes im Kriegsroman über den Ersten und Zweiten Weltkrieg* (Frankfurt: Fischer, 1984); Margrit Stickelberger-Eder, *Aufbruch 1914: Der Kriegsroman der späten Weimarer Republik* (Zurich: Artemis, 1983); Hans-Harald Müller, *Der Krieg und die Schriftsteller: Der Kriegsroman der Weimarer Republik* (Stuttgart: J. B. Metzler, 1986). For France, see the suggestive essay by Michelle Perrot, "The New Eve and the Old Adam: French Women's Condition at the Turn of the Century," in *Behind the Lines,* pp. 51–60.

90. Prost, *Anciens Combattants,* points to the phenomenon in France. Significantly, similar efforts to elevate war veterans failed in Great

Britain, see David Cannadine, "War and Death, Grief and Mourning in Modern Britain," in *Mirrors of Mortality,* ed. Joachim Whaley (New York: St. Martin's Press, 1982), pp. 187–242.

91. Alexander Kluge and Oskar Negt, *Geschichte und Eigensinn* (Frankfurt: Zweitausendundeins, 1981), pp. 797–862 with an argument that is intriguing and inscrutable in equal measure.

92. Geyer, "Krieg als Gesellschaftspolitik," esp. pp. 575–592.

Chapter 5
The Militarization of Europe, 1945–1986

1. Theo Sommer, "Ein Schritt in der falsche Richtung," *Der Zeit,* 4 April 1986.

2. Alan Bullock, *Ernest Bevin, Foreign Secretary, 1945–1951* (London: Heinemann, 1983), pp. 239–46.

3. Michael Geyer, *Deutsche Rüstungspolitik, 1860–1980* (Frankfurt am Main: Suhrkamp, 1984); Stanley Hoffmann, *Decline or Renewal? France since the 1930s* (New York: Viking, 1960), pp. 422ff.

4. Geyer *Deutsche Rüstungspolitik,* p. 192.

5. Ibid., p. 185; Hans-Peter Schwartz, *Adenauer: Der Aufstieg, 1876–1962* (Stuttgart: Deutsche Verlags-Anstalt, 1986), pp. 744, 753ff, 757.

6. Richard Barnet, *Real Security* (New York: Simon and Schuster 1981), p. 97.

7. Gordon A. Craig, "NATO and the New German Army," in *Military Policy and National Security,* ed. W. W. Kaufman (Princeton: Princeton University Press, 1956), pp. 198f.

8. Ibid, pp. 220f.

9. Ibid, pp. 225f.

10. Alan Thompson, *The Day Before Yesterday: An Illustrated History of Britain from Atlee to Macmillan* (London: Sidgwick and Jackson, 1971), pp. 200ff.

11. Henry Kissinger, *Years of Upheaval* (Boston: Little, Brown, 1982), pp. 980ff.

12. On the NATO Double-Decision, see Helga Haftendorn, *Sicherheit und Entspannung: Zur Aussenpolitik der Bundes-Republik Deutschland, 1955–1982* (Baden-Baden: Nomos, 1983), pp. 234–268.

13. See General Bernard W. Rogers, "Prescription for a Difficult Decade: The Atlantic Alliance in the 80's," *Foreign Affairs* 60 (1981–82): 1145–1156; and "Greater Flexibility for NATO's Flexible Response," *Strategic Review* (Spring 1983).

14. James Scaminaci III, "Nuclear Deterrents, Arms Control, and the Euro-Missile Crisis," Ph.D. diss. in progress, Department of Sociology, Stanford University.

15. Donald Abenheim, *Reforging the Iron Cross: The Search for Tradition in the West German Armed Forces* (Princeton: Princeton University Press, 1989). See also Gordon A. Craig, *The Germans* (New York: Oxford University Press, 1982), pp. 247–52.

16. *The Military Balance, 1985–1986* (London: International Institute for Strategical Thinking, 1986).

17. David B. Capitanchik and Richard Eickenberg, *Defense and Public Opinion* (London: Routledge and Kegan Paul, 1983).

18. Botho Strauss, *Paare Passanten,* 5th ed. (Munich: C. Hansen 1982).

19. See Fritz Raddatz, "Die Anfklarung entlässt ihre Kinder," and "Unser Verhängnis als unsere Verantwortung," in *Die Zeit,* 6, 13 July 1984.

20. Craig, *The Germans,* p. 124.

Chapter 6.
Beyond Steve Canyon and Rambo

This essay has its roots in two presentations: the first was given in March 1986 at Rutgers University's conference on "The Militarization of the World"; the second was presented in January 1987 at the conference on "Women and Military Systems," sponsored by the Peace Union of Finland. I am grateful to John Gillis of Rutgers and Eva Isaksson of the Finnish Peace Union as well as to the participants in both meetings for their encouragement, suggestions, and hunches.

1. Milton Caniff, *Steve Canyon* (Princeton: Kitchen Sink Press, 1985); Bobbie Ann Mason, *In Country* (New York: Perennial Books, 1986).

2. Cortez F. Enloe, conversation with the author, Annapolis, Maryland, February 1987; David G. Enloe, in correspondence with the

author, Seattle, Washington, February 1987; Harriett Goodridge Enloe, diaries, 1915–1982, unpublished.

3. James Fenton, "The Snap Revolution," *Granta* (London and New York: Granta, 1986); Steven D. Stark, "Ten Years into the Stallone Era," *New York Times,* 22 February 1987; Philippa Brewster, conversation with the author, London, October 1966.

4. Jean Bethke Elshtain and Sheila Tobias, eds., *Thinking about Women, Militarism and War: Essays in History, Politics and Social Criticism* (New York: Rowman and Allenheld, 1989).

5. Cynthia Enloe, *Does Khaki Become You? The Militarization of Women's Lives* (London and Boston: Pandora Press, 1988); Enloe, "Bananas, Bases and Patriarchy," *Radical America* 19, no. 4 (1985):7–23 and in Elshtain and Tobias, *Thinking about Women, Militarism and War.*

6. Elizabetta Addis, conversation with author, Cambridge, Massachusetts, November 1986.

7. Eva Isaksson, ed., *Women and Military Systems: Conference Proceedings* (Helsinki: Peace Union of Finland, 1987).

8. Claudia Koonz, *Mothers in the Fatherland: Women, the Family and Nazi Politics* (New York: St. Martin's Press, 1987); Klaus Theweleit, *Male Fantasies,* 2 vols. (Minneapolis: University of Minnesota Press, 1987) vol. 1.

9. Yumi Iwai, conversations with the author, Worcester, Massachusetts, November 1985; Kano Mikiyo, "Remolding Tennoism for Modern Japan," *AMPO,* 18, 2-3 (1986).

10. Cynthia Enloe, *Ethnic Soldiers: State Security in Divided Societies* (London: Penguin Books; Athens: University of Georgia Press, 1980).

11. Dewitt Ellinwood, ed., *India in World War I* (New Delhi: South Asia Books, 1978).

12. *The Dawn,* 5 (1889), 22 (1893), 27 (1895), 30 (1896) published in London by the British, Continental, and General Federation for the Abolition of the State Regulation of Vice.

13. Enloe, *Ethnic Soldiers.*

14. Isaksson, *Women And Military Systems.*

15. Ibid.

16. Enloe, "Bananas, Bases and Patriarchy."

17. Allan M. Brandt, *No Magic Bullet: A Social History of Venereal Disease* (New York and Oxford: Oxford University Press, 1987).

18. Ximena Bunster, "Surviving Beyond Fear: Women and Torture in Latin America," in *Women and Change in Latin America,* ed. June Nash and Helen Safa (South Hadley, Mass.: Bergin and Garvey Publishers, 1986); John Costello, *Virtue Under Fire: How World War II Changed Our Social and Sexual Attitudes* (Boston: Little Brown and Co., 1985); Kathryn Marshall, *In the Combat Zone: An Oral History of American Women in Vietnam* (Boston: Little Brown and Co., 1987); Roxanne Pastor, "How the U.S. Militarized Honduras," *Sojourner,* November 1986; Ruth Roach Pierson, *They're Still Women Afterall: The Second World War and Canadian Womanhood* (Toronto: McClelland and Stewart Publishers, 1986).

19. Paul Edwards, "The Army and the Microworld: Computers and the Militarized Politics of Gender Identity," *Signs,* forthcoming; William J. Broad, *Star Warriors* (New York: Simon and Schuster, 1987).

20. Richard Rhodes, *The Making of the Atomic Bomb* (New York: Simon and Schuster, 1986).

21. Carol Cohn, "Sex and Death in the Rational World of Defense Intellectuals," in *Thinking About Women,* ed. Elshtain and Tobias.

22. Hilary Wainwright, "Women Who Wire," in *Over Our Dead Bodies,* ed. Dorothy Thompson (London: Virago Press, 1982).

Chapter 7.
East-West versus North-South

1. For an introduction to this field of inquiry, see "The Military and the American Society," *Annals of the American Academy of Political and Social Science,* 406 (1973):1–183; Richard J. Barnet, *Roots of War* (New York: Atheneum, 1972); Alain Enthoven and Wayne K. Smith, *How Much Is Enough?* (New York: Harper & Row, 1971); Paul Y. Hammond, *Organizing for Defense: The American Military Establishment in the Twentieth Century* (Princeton: Princeton University Press, 1961); Martin B. Hickman, ed., *The Military and American Society* (Beverly Hills: Sage Publications, 1971); Samuel P. Huntington, *The Common Defense* (New York: Columbia University Press, 1971); Lyman B. Kirkpatrick, Jr., *The U.S. Intelligence Community* (New York: Hill & Wang, 1973); Seymour Melman, *The Permanent War Economy* (New York: Simon & Schuster, 1974); Harry Howe Ransome, *The Intelligence Establishment* (Cambridge, Mass.: Harvard University Press, 1970);

Marcus Raskin, "Democracy vs the National Security State," *Law and Contemporary Problems* 40 (Summer 1976):189–220; John C. Ries, *The Management of Defense* (Baltimore: Johns Hopkins University Press, 1964); and Davis Wise and Thomas Ross, *The Invisible Government* (New York: Random House, 1967).

2. On the Postwar strategic community, see Barnet, *Roots of War*, pp. 13–133; Gregg Herken, *Counsels of War* (New York: Alfred A. Knopf, 1985); Fred Kaplan, *The Wizards of Armageddon* (New York: Simon & Schuster, 1983); Peter Paret, ed., *Makers of Modern Stragegy: From Machiavelli to the Nuclear Age* (Princeton: Princeton University Press, 1986); and Bruce L. R. Smith, *The RAND Corporation* (Cambridge, Mass.: Harvard University Press, 1986).

3. See Paul Boyer, *By the Bomb's Early Light* (New York: Pantheon Books, 1985); Spencer R. Weart, *Nuclear Fear* (Cambridge, Mass.: Harvard University Press, 1988).

4. The professional literature on military and strategic affairs can best be surveyed by examining the major military and academic journals in the field, particularly *Air Force, Armed Forces Journal, Army, The Bulletin of the Atomic Scientists, Foreign Affairs, Foreign Policy, International Security, Military Review, National Defense, Naval War College Review, Orbis, Parameters, Strategic Review, Survival, U.S. Naval Institute Proceedings*, and *World Policy Journal*. Many of these journals are indexed in *Air University Library Index of Military Periodicals*. Also useful are the reports of the RAND Corporation (available in most major U.S. depository libraries), the *Adelphi Papers* and *Strategic Survey* of the International Institute for Strategic Studies (IISS) in London, and the annual *Yearbook* of the Stockholm International Peace Research Institute (SIPRI). For a comprehensive guide to the professional literature, see William M. Arkin, *Research Guide to Current Military and Strategic Affairs* (Washington: Institute for Policy Studies, 1981). Finally, for an excellent one-volume overview of these issues, see John F. Reichert and Steven R. Sturm, eds. *American Defense Policy*, 5th ed. (Baltimore: Johns Hopkins University Press, 1982).

5. For excellent accounts of the evolution of U.S. nuclear strategy, see Lawrence Freedman, *The Evolution of Nuclear Strategy* (New York: St. Martin's, 1903); Lawrence Freedman, "The First Two Generations of Nuclear Strategists," in Paret, *Makers of Modern Strategy*, pp. 735–778; Herken, *Counsels of War* and Kaplan, *Wizards*. An excellent collection of essays on the topic written by some of the key figures involved are

reprinted in Bernard F. Halloran, *Essays on Arms Control and National Security* (Washington: U.S. Arms Control and Disarmament Agency, 1986). On the air force approach, see Kaplan, *Wizards*, pp. 33–50; and David Alan Rosenberg, "The Origins of Overkill," in *Strategy and Nuclear Deterrence*, ed. Steven E. Miller (Princeton: Princeton University Press, 1984), pp. 113–181. On the theory of deterrence, see Bernard Brodie, ed., *The Absolute Weapon* (New York: Harcourt Brace, 1946); Kaplan, *Wizards*, pp. 9–32; and Glenn H. Snyder, *Deterrence and Defense* (Princeton: Princeton University Press, 1961).

6. See Kaplan, *Wizards*, pp. 74–84, 125–173; and Rosenberg, "Origins of Overkill."

7. Thomas B. Cochran, William M. Arkin, and Milton M. Hoenig, *Nuclear Weapons Databook*, vol. 1: *U.S. Nuclear Forces and Capabilities* (Cambridge, Mass.: Ballinger Press, 1984), pp. 6–13, 15.

8. See Freedman, "First Two Generations,"; Kaplan, *Wizards*, pp. 185–231, 356–391. See also Herman Kahn, *On Thermonuclear War* (Princeton: Princeton University Press, 1961); Herman Kahn, *Thinking about the Unthinkable* (New York: Horizon Press, 1962); Herman Kahn, *On Escalation* (New York: Praeger, 1965); Thomas C. Schelling, *The Strategy of Conflict* (Cambridge, Mass.: Harvard University Press, 1950); Warner R. Schilling, "U.S. Strategic Nuclear Concepts in the 1970s," in Miller, *Strategy and Nuclear Deterrence*, pp. 183–214; Richard Smoke, "The Evolution of American Defense Policy," in Reichert and Sturm, *American Defense Policy*, pp. 94–135.

9. Kaplan, *Wizards*, pp. 356–391. See also Graham T. Allison, Albert Carnesale, and Joseph S. Nye, Jr., *Hawks, Doves, and Owls* (New York: W. W. Norton, 1985); Colin S. Gray, "Nuclear Strategy: The Case for a Theory of Victory," in Miller, *Strategy and Nuclear Deterrence*, pp. 23–56; Harvard Nuclear Study Group, *Living with Nuclear Weapons* (Cambridge, Mass.: Harvard University Press, 1983); Paul Nitze, "Assuring Strategic Stability in an Era of Détente," in Halloran, *Essays*, pp. 91–120; Robert Travis Scott, ed., *The Race for Security* (Lexington, Mass.: Lexington Books, 1987); Kosta Tsipis, *Arsenal* (New York: Simon & Schuster, 1984); and Albert Wohlstetter, "Between an Unfree World and None: Increasing our Choices," *Foreign Affairs* 64 (Summer 1985):962–94.

10. The Reagan plan was first spelled out in the Department of Defense "Guidance Document" for Fiscal 1984–1985, portions of which were quoted in *The New York Times* for 1 June 1982. See also Paul

Gray, "The Reagan Nuclear Strategy," in Scott, *Race for Security,* pp. 103–115; Kaplan, *Wizards,* pp. 387–391; Christopher Paine, "Reagatomics, or How to 'Prevail,'" in Paul Joseph and Simon Rosenblum, eds. *Search for Sanity,* (Boston: South End Press, 1984), pp. 7–20; and Robert Scheer, "Pentagon Plans Aims for Victory in Nuclear War," *Los Angeles Times,* 15 August 1982.

11. See essays by Bundy, Carnesale, Kennedy, Keyworth, Pike, Rhinelander, Stares, and Drell et al. in Scott, *Race for Security,* pp. 39–94. See also Frank Blackaby, "Space Weapons and Security," in *SIPRI Yearbook 1986,* pp. 81–95; John A. Jungerman, *The Strategic Defense Initiative: A Primer and Critique* (La Jolla: Institute on Global Conflict and Cooperation, University of California at San Diego, 1988); Steven E. Miller, ed., *The Star Wars Controversy* (Princeton: Princeton University Press, 1987); John Tirman, ed., *The Fallacy of Star Wars* (New York: Random House, 1984); Union of Concerned Scientists, *Empty Promise: The Growing Case Against Star Wars* (Boston: Beacon Press, 1987).

12. See SIPRI, *Tactical Nuclear Weapons: European Perspectives* (London: Taylor and Francis, 1978); Daniel Dharels, *Nuclear Planning in NATO* (Cambridge, Mass.: Bellinger, 1987); Arthur S. Collins, Jr., "Theater Nuclear Warfare: The Battlefield," in Reichart and Sturm, *American Defense Policy,* pp. 356–365; Jonathan Dean, *Watershed in Europe* (Lexington, Mass: Lexington Books, 1987); Sverre Lodgaard and Marek Thee, eds., *Nuclear Disengagement in Europe* (London: Taylor & Francis, 1983); Jeffrey Record, *U.S. Nuclear Weapons in Europe* (Washington: Brookings Institution, 1974); Leon V. Sigal, *Nuclear Forces in Europe* (Washington: Brookings Institution, 1984).

13. For an overview of the arms control process and texts of major agreements, see Coit D. Blacker and Gloria Duffy, *International Arms Control: Issues and Agreements* (Stanford: Stanford University Press, 1984); and Jozef Goldblat, *Arms Control Agreements* (New York: Praeger, 1982). Current developments in arms control are summarized in U.S. Arms Control and Disarmament Agency, *Documents on Disarmament* (published annually); the annual *SIPRI Yearbook;* and in *Arms Control Reporter* (published monthly by the Institute for Defense and Disarmament Studies of Brookline, Mass.) For an introduction to the debate on nuclear arms control, see *Arms Control and the Arms Race: Readings from Scientific American* (New York: W. H. Freeman, 1985); Barry M. Blechman, "Do Negotiated Arms Limitations have a Future?"

in Reichert and Sturm, *American Defense Policy,* pp. 408–419; Halloran, *Essays;* Alan S. Krass, *Verification: How Much is Enough?* (London: Taylor and Francis, 1985); Scott, *Race for Security;* Strobe Talbott, *Deadly Gambits* (New York: Alfred A. Knopf, 1984); and Strobe Talbott, *Endgame: The Inside Story of Salt II* (New York: Harper & Row, 1980).

14. Kaplan, *Wizards,* pp. 263–285, 356–384.

15. For detailed information on the U.S. nuclear arsenal, see Cochran, et al., *Nuclear Weapons Databook,* vol. 1; and William M. Arkin, et al., "Nuclear Weapons," in *SIPRI Yearbook 1985,* pp. 37–80.

16. For a discussion of how American society accommodated itself to the existence of nuclear weapons, see Boyer, *By the Bomb's Early Light.*

17. For an introduction to the antinuclear literature and the text of the freeze proposal, see Joseph and Rosenbloom, *Search for Sanity.* See also Milton S. Katz, *Ban the Bomb: A History of SANE* (Westport, Conn.: Greenwood Press, 1986); and Pam Solo, *From Protest to Policy* (Cambridge, Mass.: Bellinger Press, 1988).

18. See Caspar W. Weinberger, *Annual Report to Congress, Fiscal Year 1988* (Washington, D.C.: Government Printing Office, 1987). One subsection of the report, "Force Projection and Mobilization," (26 pages out of 351) deals specifically with interventionary forces.

19. See Stephen E. Ambrose, *Rise to Globalism,* 4th rev. ed. (Harmondsworth: Penguin Books, 1985), pp. 55–78; Joyce and Gabriel Kolko, *The Limits of Power* (New York: Harper & Row, 1972), pp. 11–325; and Walter LaFeber, *America, Russia, and the Cold War, 1945–1966* (New York: John Wiley and Sons, 1968), pp. 21–36. On Kennan and containment, see "X" [George F. Kennan], "The Sources of Soviet Conduct," *Foreign Affairs* 25 (July 1947):566–82; and John Lewis Gaddis, *Strategies of Containment* (Oxford: Oxford University Press, 1982), pp. 18–53.

20. Ambrose, *Rise of Globalism,* pp. 79–98; also Richard Barnet, *Intervention and Revolution: the United States in the Third World,* rev. ed. (New York: New American Library, 1980), pp. 119–156; Richard M. Freeland, *The Truman Doctrine and the Origins of McCarthyism* (New York: Alfred A. Knopf, 1972); and Kolko and Kolko, *Limits of Power,* pp. 329–358.

21. Ambrose, *Rise of Globalism,* pp. 99–115; Gaddis, *Strategies of Containment,* pp. 54–88; Kolko and Kolko, *Limits of Power,* pp. 477–509, 534–562; and LaFeber, *America, Russia,* pp. 67–94. For the

text of NSC–68, see U.S. Department of State, *Foreign Relations of the United States, 1950* (Washington: Government Printing Office, 1977), 1:234–292.

22. Ambrose, *Rise of Globalism,* pp. 109–115; LaFeber, *America, Russia,* pp. 90–94; also Richard Freeland, *Truman Doctrine.*

23. Ambrose, *Rise to Globalism,* pp. 116–131; Gaddis, *Strategies of Containment,* pp. 89–126; Jon Halliday and Bruce Cumings, *Korea: The Unknown War* (New York: Pantheon Books, 1988); Kolko and Kolko, *Limits of Power,* pp. 565–617; LaFeber, *America, Russia,* pp. 95–122; and I. F. Stone, *The Hidden History of the Korean War,* 2d ed. (New York: Monthly Review Press, 1969).

24. Ambrose, *Rise to Globalism,* pp. 132–179; Gaddis, *Strategies of Containment,* pp. 127–197; Kolko and Kolko, *Limits of Power,* pp. 674–708; and LaFeber, *America, Russia,* pp. 123–200.

25. Maxwell D. Taylor, *The Uncertain Trumpet* (New York: Harper & Row, 1960), pp. 5–6. On flexible response and the critique of massive retaliation, see Michael Carver, "Conventional Warfare in the Nuclear Age," in Paret, *Makers of Modern Strategy,* pp. 779–789; and Michael Klare, *War Without End: American Planning for the Next Vietnam* (New York: Alfred A. Knopf, 1972), pp. 33–37.

26. On Kennedy's defense policies, see Ambrose, *Rise to Globalism,* pp. 180–200; Gaddis, *Strategies of Containment,* pp. 198–236; Roger Hilsman, *To Move a Nation* (Garden City, New York: Doubleday, 1967); Klare, *War Without End,* pp. 37–48; LaFeber, *America, Russia,* pp. 215–229; Theodore Sorenson, *Kennedy* (New York: Harper & Row, 1965); Richard J. Walton, *Cold War and Counter-Revolution: The Foreign Policy of John F. Kennedy* (New York: Viking Press, 1972). On counterinsurgency, see, Douglas S. Blaufarb, *The Counterinsurgency Era* (New York: Free Press, 1977); and John S. Pustay, *Counterinsurgency Warfare* (New York: Free Press, 1965).

27. U.S. Congress, House, Committee on Appropriations, Subcommittee, *Department of Defense Appropriations for 1964,* Hearings, 88th cong. 1st sess., 1963, Pt. 1, pp. 483–484.

28. Taylor memorandum to Robert S. McNamara, 22 January 1964, as reprinted in *The New York Times,* 13 June 1971.

29. For discussion, see Ambrose, *Rise of Globalism,* pp. 201–230; Leon Baritz, *Backfire* (New York: Ballentine Books, 1986), pp. 43–177; Gaddis, *Strategies of Containment,* pp. 237–273; George McT. Kahin, *Intervention: How America Became Involved in Vietnam* (New York:

Alfred A. Knopf, 1986), pp. 146–235; Stanley Karnow, *Vietnam, A History* (Harmondsworth: Penguin, 1984), pp. 312–473; and Gabriel Kolko, *Anatomy of a War* (New York: Pantheon Books, 1985).

30. Baritz, *Backfire,* pp. 178–222; Karnow, *Vietnam,* pp. 473–612; also Todd Gitlin, *The Sixties: Years of Hope, Days of Rage* (New York: Bantam Books, 1987).

31. Gaddis, *Strategies of Containment,* pp. 274–308; and Eliot A. Cohen, "Constraints on America's Conduct of Small Wars," in *Conventional Forces and American Defense Policy,* ed. Steven E. Miller (Princeton: Princeton University Press, 1986), pp. 277–308.

32. Ambrose, *Rise of Globalism,* pp. 301–321; Michael T. Klare, *Beyond the 'Vietnam Syndrome'* (Washington, D.C.: Institute for Policy Studies, 1981), pp. 1–14; and Gabriel Kolko, *Confronting the Third World* (New York: Pantheon Books, 1988), pp. 205–298.

33. Reagan's comments on the Vietnam syndrome were made in his commencement address at the U.S. Military Academy at West Point, New York, 27 May 1981 and reprinted in *The New York Times,* 28 May 1981. For discussion, see Ambrose, *Rise to Globalism,* pp. 322–345; Noam Chomsky, *Toward a New Cold War* (New York: Pantheon Books, 1982), pp. 216–229; and Barry R. Posen and Steve Van Evera, "Defense Policy and the Reagan Administration: Departure from Containment," in Miller, *Conventional Forces,* pp. 63–78. On the build-up of power projection forces, see Stephen D. Goose, "Low-Intensity Warfare: The Warriors and Their Weapons," in *Low-Intensity Warfare,* ed. Michael T. Klare and Peter Kornbluh (New York: Pantheon Books, 1988), pp. 80–111.

34. On LIC doctrine, see Michael Klare, "The Interventionist Impulse: U.S. Military Doctrine for Low-Intensity Warfare," in *Low-Intensity Warfare,* pp. 49–79; Donald R. Morelli and Michael M. Ferguson, "Low-Intensity Conflict: An Operational Perspective," *Military Review* 64 (November 1984): 2–16; Sam C. Sarkesian, "Low-Intensity Conflict: Concepts, Principles, and Policy Guidelines," *Air University Review* (January-February 1985): 1–13; Robert J. Ward, "LIC Strategy," *Military Intelligence* (January-March 1985): 52–60; and Weinberger, *Annual Report,* Fiscal 1988, pp. 56–62. On El Salvador, see Daniel Seigel and Joy Hackel, "El Salvador: Counterinsurgency Revisted," in Klare and Kornbluh, *Low-Intensity Warfare,* pp. 112–135.

35. On the Reagan Doctrine, see Richard L. Armitage, "Tackling the Thorny Questions on Anti-Communist Insurgencies," *Defense/85*

(October 1985): 15–20; William R. Bode, "The Reagan Doctrine," *Strategic Review* 14 (Winter 1986): 21–29; Raymond W. Copson and Richard P. Cronin, "The 'Reagan Doctrine' and its Prospects," *Survival* 29 (January-February 1987): 40–55; and Peter Kornbluh, "Nicaragua: U.S. Proinsurgency Warfare Against the Sandinistas," in Klare and Kornbluh, *Low-Intensity Warfare*, pp. 136–157. On the Iran-Contra affair, see *The Tower Commission Report*, Report of the President's Special Review Board (New York: Bantam and Times Books, 1987); and *Report of the Congressional Committees Investigating the Iran-Contra Affair* (Washington, D.C.: Government Printing Office, 1987).

36. On the psychological impact of nuclear weapons, see Robert J. Lifton, "Beyond Psychic Numbing: A Call for Awareness," *Journal of Orthopsychiatry* 52 (1984):619–629; Weart, *Nuclear Fear;* and *"Images of Nuclear War,"* special issue of *Journal of Social Issues* 39 (1983).

37. Marshall D. Shulman, *East-West Tensions in the Third World* (New York: W. W. Norton, 1986), p. 5.

38. Walter LeFeber, *Inevitable Revolutions: The United States in Central America* (New York: W. W. Norton, 1983), pp. 19–83.

39. For discussion, see Barnet, *Intervention and Revolution*, pp. 13–115; Jonathan Kwitney, *Endless Enemies* (Harmondsworth: Penguin Books, 1986); and Kolko, *Confronting the Third World*.

40. See polling data and anlaysis by Martill and Kiley Agency of Boston in *Turnabout: The Emerging New Realism in the Nuclear Age* (Boston: Women's Action for Nuclear Disarmament, 1986).

41. In 1987 a poll of American voters conducted by Mellman and Lazarus Research of Washington, D.C. found that 48 percent of the electorate considered economic disorder as the greatest threat to U.S. security, 31 percent cited terrorism, and only 17 percent cited the Soviet military threat. See "Defining American Strength," a survey conducted for the World Policy Institute, 15–20 October 1987.

42. Guy Pauker, *Military Implications of a Possible World Order Crisis in the 1980s*, Report No. R–2003–AF (Santa Monica: Rand Corp., 1977), pp. 1–4.

43. Neil C. Livingstone, "Fighting Terrorism and 'Dirty Little Wars,'" in *Defense Planning for the 1990s*, ed. William A. Buckingham, Jr. (Washington: National Defense University Press, 1984), pp. 166, 186.

44. James B. Motley, "A Perspective on Low-Intensity Conflict," *Military Review* (January 1985):9.

45. *Discriminate Deterrence*, Report of the U.S. Commission on Inte-

grated Long-Term Strategy (Washington, D.C. 1988), pp. 13–14, 33–34.

46. On the arms trade with the Third World, see Michael Brzoska and Thoman Ohlson, *Arms Transfers in the Third World 1971–85* (Oxford: Oxford University Press, 1987); Michael Klare, *American Arms Supermarket* (Austin: University of Texas Press, 1985); and Andrew Pierre, *The Global Politics of Arms Sales* (Princeton: Princeton University Press, 1982).

47. Maxwell D. Taylor, "The Legitimate Claims of National Security," *Foreign Affairs* 52 (April 1974):586.

48. See Michael Geyer's contribution to this volume.

Index